BUILDING CONSTRUCTION
METHODS AND MATERIALS
FOR THE FIRE SERVICE
Second Edition

Michael Smith

PEARSON

Boston Columbus Indianapolis New York San Francisco Upper Saddle River Amsterdam
Cape Town Dubai London Madrid Milan Munich Paris Montreal Toronto Delhi
Mexico City São Paulo Sydney Hong Kong Seoul Singapore Taipei Tokyo

Publisher: Julie Levin Alexander
Publisher's Assistant: Regina Bruno
Senior Acquisitions Editor: Stephen Smith
Associate Editor: Monica Moosang
Development Editor: Donna Morrissey
Editorial Assistant: Samantha Sheehan
Director of Marketing: David Gesell
Marketing Manager: Brian Hoehl
Marketing Specialist: Michael Sirinides
Marketing Assistant: Crystal Gonzalez
Managing Production Editor: Patrick Walsh
Production Liaison: Julie Boddorf
Production Editor: Lisa Garboski, bookworks
Senior Media Editor: Amy Peltier
Media Project Manager: Lorena Cerisano
Manufacturing Manager: Alan Fischer
Creative Director: Jayne Conte
Cover Designer: Bruce Kenselaar
Cover Photo: Harvey Eisner
Composition: S4Carlisle Publishing Services
Printing and Binding: R.R. Donnelley/Willard
Cover Printer: Lehigh-Phoenix Color/Hagerstown

Credits and acknowledgments borrowed from other sources and reproduced, with permission, in this textbook appear on appropriate pages within the text. Unless otherwise stated, all photos have been provided by the authors.

Library of Congress Cataloging-in-Publication Data
Smith, Michael (Michael Lewis)
    Building construction: methods and materials for the fire service/Michael Smith. -- 2nd ed.
    ISBN 0-13-708378-5
1. Building. 2. Buildings. 3. Building materials. 4. Fire prevention.
5. Building failures. 6. Structural engineering. I. Title.
    TH153.S733 2012
    628.92—dc23
                        2011014754

Pearson® is a registered trademark of Pearson PLC.

10 9 8 7 6 5 4 3 2 1

ISBN 10:   0-13-708378-5
ISBN 13: 978-0-13-708378-7

CONTENTS

Chapter 3 Building Materials: Stone, Masonry, Steel, Concrete, and Wood 47

Chapter 4 Building Components 68

Chapter 5 Building Systems 100

Chapter 6 Building Construction Types 132

FOREWORD

I am honored to write this foreword for my good friend, Deputy Chief Michael Smith (retired). He was a Deputy Chief with the Washington D.C. Fire Department, retiring after thirty-one years, and he had a career as a master builder. Not only has he known the dangers of crawling down smoke-filled hallways searching burning structures for signs of life, but he also labored to construct buildings. Chief Smith's innate knowledge of the many nuances of the various types of construction that we, as firefighters, are daily called to fight fires in, is enhanced by his many years of laboring in the building construction industry. There is no doubt that firefighting and building construction are interwoven and the fact that Chief Smith is sharing that knowledge with firefighters is beneficial to everyone.

Chief Smith's first edition of *Building Construction: Methods and Materials for the Fire Service* was a book written by a firefighter who had been there and done that. Mike Smith is a safety-conscious individual who is truly interested in providing information that is both educational and useful for all firefighters. Others have written books on building construction, but none have the combination of firefighting and building construction experience of Chief Smith. His book gives us insight into all aspects of building construction. As excellent as the first edition was, this second edition delves deeper into the knowledge that we need for safe incident scene operations.

One might ask "How important is it for a firefighter to have knowledge of building construction?" Firefighters need to realize that it would be practically impossible to predict fire travel within a structure without an understanding of the building's construction. It is important to know what vertical openings normally exist that could permit fire travel. Are there open stairs as typically found in a multistory single-family dwelling or are the vertical openings protected by self-closing doors as found in other occupancies? Another factor is that the potential for building collapse is accentuated in buildings using lightweight forms of construction as illustrated by the firefighter injuries and deaths that have occurred in these structures. Information on all types of construction is contained within this book.

Building Construction: Methods and Materials for the Fire Service's second edition has a new look and new features. One addition titled Firehouse Discussion examines significant incidents that underscore the role that a building and its systems play in the outcome of an incident. Chief Smith's introspective look allows the reader to comprehend not only what has occurred, but prepares us as firefighters and students of building construction to apply that knowledge if presented with similar circumstances.

First Response is another new feature that illuminates various building occupancies and offers some tactical and strategically-based considerations for the responding firefighters as well as the company and chief officers. First Response allows us to gain the firefighting insight as applied to various occupancies from Chief Smith's professional perspective.

Building Green is another new feature that recognizes that some of the most significant advances in the building industry are related to going green. From lightweight buildings that reduce the size and amount of wood and other natural resources used in construction, to the installation of rooftop solar panels, Building Green looks at the impact these changes can have on firefighting operations, from the increased potential for building collapse to the inability to perform ventilation on a rooftop that is fully covered with solar panels.

There are many added extras in the second edition of *Building Construction: Methods and Materials for the Fire Service*. All the chapters have been updated and expanded to include valuable information regarding the building industry, such as elevator systems without machine rooms and windows and doors installed to thwart terrorism. The chapter on trusses has been expanded and now includes other manufactured products and assemblies. Also included are additional color visuals to illustrate what Chief Smith believes is important information for every responder.

James P. Smith
Deputy Chief (Retired),
Philadelphia Fire Department

Introduction

I always wanted to be a fireman (now, to be politically correct, referred to as firefighter). I grew up in West Philadelphia and Clifton Heights, Pennsylvania. After I got out of the service, I began my journey toward becoming a member of a fire department. I tested in twenty cities across the country. Finally, after a year, I was appointed to the District of Columbia Fire Department in Washington, D.C. I had the distinct privilege of working with some great firefighters. It was my profession, so even if it wasn't was milk and honey all of the time, I would not have traded it for anything.

After thirty-one years, I've now completed my career and that is why I wrote this book. During those years, thirteen members of my department were killed in the line of duty. Over 3,500 firefighters were killed across the country in line-of-duty–connected incidents; however, not all of them had to die. Some time ago I wrote an article in which I said we are not finding new ways to kill our firefighters; we are desperately trying to perfect the old ways. I cringe when members of a department fall through a floor, are struck from above by failing walls or roofs, are incinerated in 19′ × 33′ two-story rowhomes, or get lost in commercial or high-rise structures. Unfortunately, even if these firefighters had been searching because of the ultimate reason that we all serve—to save a life—none of these people were found with victims in their arms. In many cases, the victims were dead before the department arrived.

The incidents have changed since I first put on a set of turnouts. The structures built in the last forty years are more lightweight and filled with plastics and other highly combustible materials. The fires occurring today are hotter and less forgiving than when I went down the hallway. We are better encapsulated, which appears to give a false sense of security. So more and more we are forced to read about firefighters killed in truss collapses or horribly burned in flashovers.

Our knowledge of building construction is woefully inadequate and a rookie can no longer rely on the experience of the veteran because the vet may not have been to that many incidents. In my early years we went to a lot of jobs, fires where water was pumped through an attack line. On many occasions we were taking up from one fire only to respond immediately to another. It was not unusual to go to five or six working incidents in a ten- or fourteen-hour tour. Most of the critiques and instructions were given by the senior members on the rig. I had a captain who told me not to worry about screwing up as I would get another chance by the end of the tour of duty. His emphasis was on tool competency and learning the buildings in our response area.

Training has diminished. When I started as a volunteer firefighter in Clifton Heights, training was available every weekend: live fire training involving structures and flammable liquids. And we were required to go. Yes, required, or we didn't ride. Training was never designed to be an initiation or to test courage. Today we can't have live training fires because students are too often killed. Strict compliance with NFPA 1403: *Standard on Live Fire Training Evolutions* should be the goal of every fire service instructor. As instructors, however, we need to set the example of safety, the concept of teamwork, and the need for recognizing the enemy and its capabilities. This recognition is called size-up and as a service we don't do it well. Begin reading the NIOSH after-action reports at www.cdc.gov/niosh/fire and too often you find the words *routine* and *size-up* together. Firefighters run into disasters that can be predicted. I've wondered if they were running because they truly believed they could be successful or if they just didn't know the dangers.

This book is intended to assist in correcting these issues. It is designed to open the door to the wonderful world of construction. I also worked for over thirty years in the construction industry as a carpenter, foreman, and finally as superintendent. I continue to work in that trade, helping volunteer agencies assist the homeless and veterans. I continue to bang nails, not for a living anymore, but to try to give something back. So I am still part of the construction process. Nothing in this book is intended to cast dispersions on the men and women of that industry; however, their efforts are to thwart gravity, not fire or natural disasters or human-made incidents. We in the fire service need to learn more about buildings and how they are put together so that when we respond, we will understand how and when they come apart.

Some readers of the first edition and early readers of this, the second edition, questioned why is there a need to study codes or plans. The answer is that we need to move past the mantra that only by quickly putting the wet stuff on the red stuff will we do our jobs. Knowledge of codes and plans allows you to understand the frailties of structures during a fire and also what is really behind that exterior covering of a building. This book is not intended to be used for one semester or to study for promotion only. It is not intended for one segment of the firefighting work force. It is not intended for brand new recruits only.

It is intended to be a useful tool to be used during inspections, EMS calls, visits to a construction site during lunch, or any time that curiosity dictates that the user wants to learn more about the places where they might have to ply their trade. It is also intended to be used by all ranks and firefighters of all experience levels. If there is something that you, as the reader, do not understand, then ask someone who does. Ideally it will be your instructor or one of your more seasoned co-workers. If all else fails, contact me through the publisher and I will be happy to explain.

One reviewer questioned why the steel-making process is included in the book. It is included because we as firefighters do not appreciate the vulnerability of steel when exposed to fire. The same holds true for trusses. Firefighters are being killed in more buildings that contain trusses then ever before and yet we still have those who are proponents of aggressive interior attacks when no life is confirmed to be at risk other than those of the firefighters sent into an ambush.

Organization

Terms such as *masonry* and *ordinary construction* are used often, but too many of us don't really understand what those terms imply. Chapters 1–7 are designed to take you through the methods and materials of construction. Chapters 1 and 2 discuss forces and their implications and how codes are constructed to handle these forces. Chapter 3 presents the principal materials used in building. Information on how these materials are made and installed is included to help you identify what you are really looking at during the size-up phase of the response. Chapter 4 gives you an understanding of the components of a building and how they interact with one another. Chapter 5 discusses electrical, plumbing, HVAC, and elevator systems. Chapter 6 is designed to help you identify building types by identifiable cues. Chapter 7 covers trusses and other manufactured and engineered systems, both metal and wood. Chapter 8 focuses on size-up cues regarding construction and preparedness. Chapter 9 was written to help you identify the potential for failures by building type. Finally, Chapter 10 addresses commanding a structural collapse incident and underscores many of the unique issues for responses of this type.

Features

- *Key Terms* lists appear at the beginning of each chapter and key terms are defined in the margins where first introduced. Full definitions are provided in the comprehensive glossary.
- *Stay Safe* icons within each chapter provide important lifesaving points, calling attention to situations or conditions that could severely impact firefighters.

- *Firehouse Discussion* and *Think About It* are features that present synopses of fires in which the building had a significant impact on the fatalities. Some of these involve civilian fatalities while others involve the deaths of firefighters.
- *The School of Hard Knocks* features include real-world incidents experienced by the author. This is lessons-learned content that allows you to learn best practices and avoid future mistakes.
- *On Scene* case studies, including critical thinking questions, appear in each chapter and are designed to reinforce key chapter concepts.
- *First Response* content provides concise tactical and strategic considerations for various occupancies encountered by firefighters.
- *Review Questions* appear at the end of each chapter, prompting you to test yourself on what has been presented.
- *Suggested Readings* are listed at the end of each chapter, providing you with additional resources for acquiring further knowledge.

At the completion of selected sections, we prompt readers to visit Resource Central on www.bradybooks.com to view additional information on topics discussed. Students will also find quizzes, additional web links, and more to supplement classroom learning. Through Resource Central, this text also offers instructors a full complement of online supplemental teaching materials such as test banks and PowerPoint lectures to aid in the classroom.

This book is committed to providing information to help you better understand the consequences of building processes on the fire service. My hope is that after reading this book you will become a student of building processes as I have and make studying it a major part of your career development. Stay safe!

Michael Smith

ACKNOWLEDGMENTS

I have been extremely fortunate over the years. I have had mentors and friends who have made many of my successes possible.

First, I have to recognize the fire service mentors who answered my questions and inspired me to continue along my path when it would have been easier politically to cave in and go along to get along. Not only were they mentors and coaches, but they gave me friendship as well. In my opinion, they are truly the masters of the work that we do. They are Leo Stapleton, Vinny Dunn, Jim Smith, Denis Onieal, Charlie Dickinson, Harvey Eisner, and the late Howie McClennan.

I had mentors and coaches at the DCFD as well. Robert McIlwraith, Howard Dixon, Charles Culver, Gamelia Jackson, Alexander Bullock, and William Mullikin brought me truth, gave me the history of how and why fire departments evolve, and spent countless hours trying to guide me along the path of becoming a better firefighter. At times, I wish I had listened more. There are countless others whose words and deeds inspired me to be more than I ever anticipated.

My career in the construction industry contained mentors and coaches also. These men had a profound effect on my appreciation for the process of building. Melvin Paige, Charley Butler, Bob Gruver, Alex "Sandy" Gourlay, Ray Askins, and the late Jim Blandford took me from apprentice to master over the period of twenty-five years.

I want to recognize my first fire chiefs: John Gorman and Ed Volante set the standard for being a chief. I will always be grateful for their motivation, mentoring, and encouragement. I will always be grateful to all of the volunteers who still man the rigs at the Clifton Heights Fire Company; they are still doing it the right way and they make me proud to be associated with them.

I have made many friends in the fire service during my tenure. I will always be grateful for the camaraderie, the friendship, and the brotherhood. In a time when many espouse the term "brotherhood," the men and women who have been my friends live it every day without question.

Completing a second edition would appear to be simple; after all, the book is written already. But it's not as easy as it appears. I will be forever grateful to the reviewers for their comments; a full list appears below. The team at Pearson has been instrumental to whatever success that this edition will have. My development editor, Donna L. Morrissey, kept me on the right path in a timely fashion. Stephen Smith supported this project from the beginning; I hope this edition makes him proud. I can never thank Monica Moosang enough for her behind-the-scenes efforts and critical input. There are many others who will forever be in my thoughts.

I want to thank all the men and women, volunteer and paid, who still go down the hall to fight the beast. Your courage and dedication to duty is inspiring. I hope the information within this book will help to keep you safe. Your world is far more dangerous than mine ever was. Stay Safe!

The publisher and author would like to extend their thanks to the following reviewers, whose feedback helped to shape the final text.

Second edition:

Capt. Joseph V. Bruni, Jr.
St. Petersburg Fire & Rescue
Adjunct instructor, St. Petersburg
  College
St. Petersburg, FL

Andrew Byrnes, M.Ed. EFO
Utah Valley University
Orem, UT

Lt. John Caran, AAS Fire Science,
EMT-I
Adjunct Instructor, Augusta Technical
  College
Adjunct Instructor, Georgia Fire
  Academy
Augusta, GA

Gary W. Edwards
MSU Billings–College of Technology
Fire Science Program
Billings, MT

J. L. Journeay, M.A.
CSU College of Safety and Emergency
  Services
Chair, Department of Fire Science
Orange Beach, AL

Chief Joe Mercieri
Littleton Fire Department
Littleton, NH

Adam D. Piskura
Director, Connecticut Fire Academy
Windsor Locks, CT

Paul Reynolds, AAS Fire Protection,
B.S. Training and Education
Southwestern Oregon Community
  College Fire Science Branch
Coos Bay, OR

Douglas E. Rohn
Madison Fire Department
Madison, WI

Nathan Sivils
Director, Fire Science
Blinn College
Bryan, TX

Martin Walsh, C.F.P.S., B.S.
San Diego Miramar College
San Diego, CA

Al Wickline
Adjunct Faculty
Allegheny County Fire Administrator
Monroeville, PA

First edition:

Attila Hertelendy
Mississippi State Fire Academy
Jackson, MS

John Eric Pearce
Rio Hondo College
Whittier, CA

David Walsh
Program Chairperson
Dutchess Community College
Poughkeepsie, NY

Dale Anderson
Director of Fire Science
Casper College
Casper, WY

ABOUT THE AUTHOR

Michael Smith worked with the fire service for over thirty-five years, retiring with the rank of Deputy Fire Chief. He has also worked for over thirty years in the construction industry; very few firefighters made enough money to work only one job.

He holds an A.S. degree in Fire Science; a B.S. degree in Construction Management; and an M.A. degree in Fire and Public Administration. He is a graduate of the Executive Fire Officers Program at the National Fire Academy. He is also a Certified Municipal Manager from George Washington University.

He is currently involved with lecturing, consulting, and emergency management for the federal government, the fire service, and the United States military. He can be contacted through the publisher.

Building Construction: Methods and Materials for the Fire Service, Second edition

By Michael Smith

The second edition has a new 4/color design, and many new features, including:

- *Firehouse Discussion* features examine significant incidents that underscore the role that a building and its systems play on the outcome of an incident. Chief Smith's introspective look allows the reader to comprehend not only what has occurred, but prepares firefighters and students of building construction to apply that knowledge, if presented with similar circumstances.
- *First Response* features illuminate various building occupancies and offers some tactical and strategically based considerations for the responding firefighters as well as the company and chief officers.
- *Building Green* features recognize that some of the most significant advances in the building industry are "going green." Looks at the impact these changes can have on firefighting operations, from the increased potential for building collapse to the inability to perform ventilation on a rooftop that is fully covered with solar panels.
- New *Resource Central* solution offers unique book-specific online resources for both students and instructors in a user-friendly format, including review questions and chapter reinforcement materials for the student and Instructor's Manual, PowerPoint, Online Courseware, and Test Banks for the instructor.
- *CourseSmart ebook* option
- *Full Online Courseware* available

Updated and expanded, the second edition offers many added extras.

- All chapters have been updated and expanded to include valuable information regarding the building industry, such as elevator systems without machine rooms and windows and doors installed to thwart terrorism.
- The chapter on trusses has been expanded and now includes other manufactured products and assemblies.
- The addition of color visuals help illustrate what Chief Smith believes is important information for every responder.

Fire and Emergency Services Higher Education (FESHE) Grid

The following grid outlines Building Construction course requirements developed as part of the FESHE Model Curriculum and where specific content can be located within this text:

Course Requirements	1	2	3	4	5	6	7	8	9	10
Describe building construction as it relates to firefighter safety, building codes, fire prevention, code inspection, firefighting strategy, and tactics.	X	X	X	X	X	X	X	X	X	X
Classify major types of building construction in accordance with local/model building codes.		X				X		X	X	X
Analyze the hazards and tactical considerations associated with various types of building construction.				X	X	X	X	X	X	X
Explain the different loads and stresses that are placed on a building and their interrelationships.	X								X	
Identify the function of each principal structural component in typical building design.			X	X		X				
Differentiate between fire resistance and flame spread; describe the testing procedures used to establish ratings for each.		X								
Classify occupancy designations of building codes.		X								
Identify the indicators of potential structural failure as they relate to firefighter safety.				X			X	X	X	X
Identify the role of GIS as it relates to building construction.		X								

Fire and Emergency Services Higher Education (FESHE) Grid

The following grid outlines the titles, content, and course requirements developed as part of the FESHE Model Curriculum and where specific content can be located within this text.

Loads and Other Forces

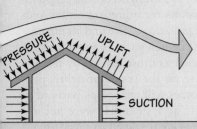

OBJECTIVES

After reading this chapter, you should be able to:

■ Understand the significance of loads on a structure.
■ Define key loads and how they affect structural integrity.
■ Understand the relevance of identifying how structural members are located in the building.

Introduction

This chapter examines how various forces affect structural components. These forces are presented as *loads*. The position of the structural member, whether vertical or horizontal, and its place in the overall system of the structure, can have a profound impact on the safety of the occupants within the structure as well as that of the firefighters who respond to the incident. Chief officers cannot make informed decisions regarding strategies and tactics unless they understand how these forces react to the presence of fire and its effects. Units cannot respond to a structure collapse from natural causes such as earthquakes, hurricanes, tornadoes, floods, or human-made collapses, such as those caused by terrorism, unless they understand the effects that forces will have on the resultant debris pile or remnants of the structure. This chapter looks at how forces will change as parts of a system begin to fail and loads shift to other parts of a building system that were not intended to carry these loads. It also looks at how many systems are designed to carry forces in the same manner, systems that may possess the same limitations even though they are dissimilar in material composition. For example, concrete and trusses are excellent for carrying loads in **compression** but are very poor at carrying loads in **tension**.

Humans have fought the forces of mother nature ever since deciding they didn't want to live exposed to the elements. Early humans found shelter in caves. When the first settlers arrived in this country they took what was available in nature to construct their shelters. The east coast offered wood and stone, the prairies offered sod, the far west offered a plethora of materials including mud, stone, and wood.

Dead and Live Loads

As soon as people raise up a structure they encounter *gravity*. Gravity is a natural force, exerted twenty-four hours a day, seven days a week. It is caused by the gravitational pull of the earth and acts in a vertical direction. This force pulls down on everything, including structures. The force increases for every inch that a structure rises above the ground. Two types of loads imposed by gravity are **dead loads** and **live loads** (see Figure 1-1). A dead load is the building itself with all its structural components and the permanently affixed ancillary items, including HVAC units, fire escapes, plumbing and electrical equipment, walls, doors, and windows.

compression ■ Squeezing together; making smaller by direct pressure.

tension ■ A pulling or stretching force, the opposite of compression, which causes a crushing strain.

dead loads ■ Permanent, inert loads whose pressure on a building is steady and constant due to the weight of its structural members and the fixed loads they carry; these impose definite loads and strains.

live loads ■ The moving load or variable weight to which a building is subjected, due to the combined weight of the people who occupy it, the furnishings, and other movable objects.

FIREHOUSE DISCUSSION

The Vendome Hotel site is located in Boston at the intersection of Commonwealth Avenue and Dartmouth Street. The hotel was first constructed in 1871 and was expanded in 1881. In 1890, a fateful decision was made to construct a ballroom on the first floor. This renovation involved removing the main load-bearing wall on the first floor. The only remaining support for the upper five floors was a single 7″ diameter cast iron column. This situation remained innocuous until June 17, 1972. The hotel had been sold after remaining vacant for several years. The intent was to convert the Vendome into condominiums and a shopping mall. On the day of the fire, workers were the only occupants. A fire was discovered by the workers at 2:35 PM in an enclosed space between the third and fourth floors. The Boston Fire Department was called and by 3:06 PM a total of four alarms had been struck, bringing sixteen engines, five ladders, two aerial towers, and a heavy rescue to the hotel. A fire attack force of over 100 firefighters fought the blaze for over two hours bringing the fire under control at approximately 4:30 PM. A combination of 2½″ handlines and ladder pipes had been employed during the battle. Overhaul, the process of digging out the interior structure to determine the extent of the spread of the fire and to quell any remaining fire, was begun by the members of Ladder 13 and Engines 22 and 32. At 5:28 PM all five floors of a 40′ × 45′ section at the southeast corner of the building collapsed, burying seventeen firefighters under a two-story pile of debris. Nine firefighters lost their lives. The hotel site is now comprised of 110 residential condominium units and 27 commercial units.

THINK ABOUT IT!

Assuming that multiple ladder pipes and large-caliber handlines were deployed at the height of the battle, 1000 to 4000 gallons per minute (GPM) were being delivered inside the structure. Using the usual fire service value of 8.67 pounds per gallon, and assuming a factor of 35 percent of the water being converted directly to steam (this is not exact science, but has been used successfully when employing larger streams) and a ratio of 1:3 for water leaving the building, the result means that for every 3 gallons of water delivered, minus the steam conversion, 1 gallon is running out of the building.

- Approximately how much weight would be applied to each floor?
- As the water is applied and then moves throughout the building, what forces is it applying to the structural components?
- After sustaining a large caliber attack, would it be prudent to allow the building to remain unoccupied until all water has drained from the building?
- Should building engineers be called to the scene to declare the building tenable for subsequent occupancy by the fire inspectors and crews performing overhaul?
- What role should a fire department safety officer play in a response such as this and how many personnel would you assign to safety?

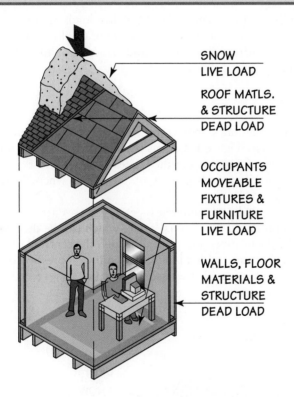

SNOW
LIVE LOAD

ROOF MATLS.
& STRUCTURE
DEAD LOAD

OCCUPANTS
MOVEABLE
FIXTURES &
FURNITURE
LIVE LOAD

WALLS, FLOOR
MATERIALS &
STRUCTURE
DEAD LOAD

FIGURE 1-1 Typical building loads. *Source: Onouye, Barry S.; Kane, Kevin,* Statics and Strength of Materials in Architecture and Building Construction, *2nd Edition, © 2002, p. 7. Reprinted by permission of Pearson Education, Inc., Upper Saddle River, NJ.*

If the structure is more than one story, all loads above a particular building component are also added to the designed dead load. The bottom steel columns of the Empire State Building in New York City were designed to carry 10,000,000 pounds each.

Over the years, humans have developed lighter-weight materials and components able to carry the same loads as their heavier predecessors. Trusses, both wood and metal, are one example of this evolution. As dead loads have decreased, structural materials have become lighter and also less redundant, but are thus, using the theory that less mass causes more effect, quicker to be affected by fire.

Live loads include furniture, people, movable equipment, or stored materials. Live loads also include snow and rain, but not wind or earthquakes. Live loads may vary from

TABLE 1-1	Minimum Design Loads for Buildings and Other Structures		
OCCUPANCY OR USE	**LIVE LOAD (PSF)**	**OCCUPANCY OR USE**	**LIVE LOAD (PSF)**
Assembly Areas and Theaters		Office Buildings	
Fixed Seats	60	Lobbies	100
Movable Seats	100	Offices	50
Dance Halls and Ballrooms	100	Residential	
Garages (passenger cars only)	50	Dwellings (one- and two-family)	40
Storage Warehouse		Hotels and Multifamily Houses	40
Light	125	School Classrooms	40
Heavy	250		

Source: Merritt, Frederick S. (1996). "Minimum Design Loads for Buildings and Other Structures: An American Society of Civil Engineers Standard 7-95." Journal of Architectural Engineering. Used with permission from ASCE.

floor to floor depending on the occupancy. Most live loads are determined by a method specified by local jurisdiction. The American Society of Civil Engineers (ASCE) has developed specifications for various occupancies and structures and many jurisdictions have adopted ASCE standards (see Table 1-1.)

Stress

When a force acts on a building system component member, the member will react to the applied force internally (see Figure 1-2). This reaction to the applied force is called *stress*. The most common method engineers in the United States use to express stress is pounds per square inch (psi) or kilopounds per square inch (ksi), where 1 kilopound (kip K) = 1000 pounds (see Figure 1-3).

The two most commonly encountered types of stresses are axial stress and flexural stress. **Axial** stress is produced by an axial force, which can be defined as a force that is parallel to the axis of the member and passes through the centroid of the cross section of the member. A vertical column with a load imposed in compression directly on top of its vertical axis is an example of axial stress (see Figure 1-4).

axial ■ Situated in or on an axis.

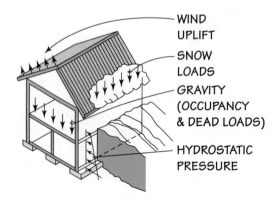

FIGURE 1-2 Array of forces acting on a house at the same time. *Source: Onouye, Barry S.; Kane, Kevin, Statics and Strength of Materials in Architecture and Building Construction, 2nd Edition, © 2002, p. 26. Reprinted by permission of Pearson Education, Inc., Upper Saddle River, NJ.*

WIND
UPLIFT

SNOW
LOADS

GRAVITY
(OCCUPANCY
& DEAD LOADS)

HYDROSTATIC
PRESSURE

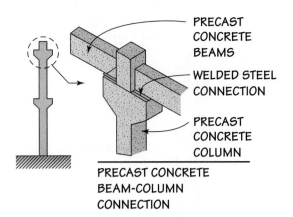

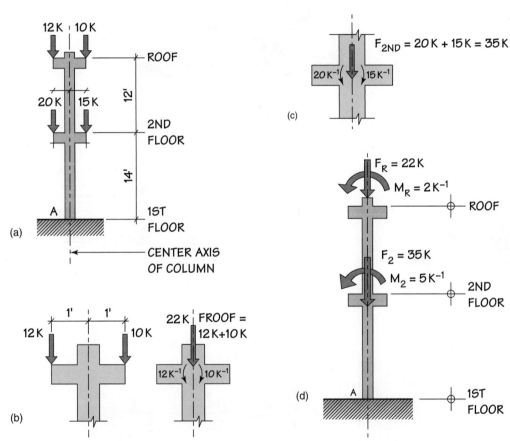

FIGURE 1-3 Example of loads through connection, using kips to express force. *Source: Onouye, Barry S.; Kane, Kevin, Statics and Strength of Materials in Architecture and Building Construction, 2nd Edition, © 2002, p. 75. Reprinted by permission of Pearson Education, Inc., Upper Saddle River, NJ.*

Flexural stresses are incurred when the applied force is perpendicular to the axis of the member. These can also be referred to as *bending* stresses (see Figure 1-5). When a force is applied parallel to the axis of a member but does not pass through the center of the cross section of that member, the stress is referred to as being *eccentric*. A point loading where the weight of the load is resting on a specific point of a floor area adjacent to a column is said to be eccentric to that column.

FIGURE 1-4 A column supporting an external load in axial stress. *Source: Onouye, Barry S.; Kane, Kevin, Statics and Strength of Materials in Architecture and Building Construction, 2nd Edition, © 2002, p. 23. Reprinted by permission of Pearson Education, Inc., Upper Saddle River, NJ.*

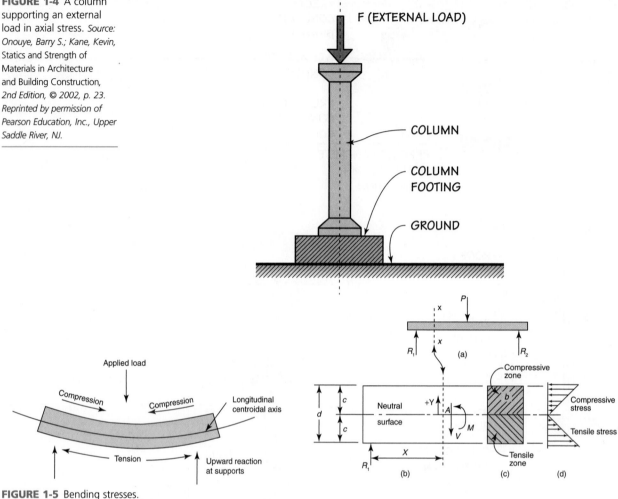

F (EXTERNAL LOAD)

COLUMN

COLUMN FOOTING

GROUND

FIGURE 1-5 Bending stresses.

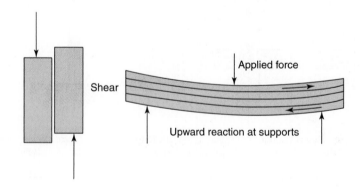

FIGURE 1-6 Example of deformation showing compression and tension forces.

DEFORMATION

Deformation accompanies stress. *Deformation* can be defined as the change in the size and shape of a member under load. Tensile stress will cause the member to elongate and compressive stress will cause the member to shorten (see Figure 1-6).

The resultant change in the length of the member is referred to as *strain*. Normally the strain will increase in proportion to the stress. Under normal conditions the deformation of a member under load is imperceptible to the naked eye. This is because most building components are considered to be rigid. The strength of a material depends on its ability to sustain a load without undue deformation or failure. This value can be derived by experiment. The American Society for Testing and Materials (ASTM) has published guidelines and limits for conducting tests and for measuring the resultant strength, with limits for standards for the specific applications that are acceptable for the material being tested. One of the most widely used tests is the compression–tension test. For this test, a sample is provided in the designated shape of the desired end product. A marking system of dots is sometimes used and the sample is placed in the position for testing. The ends of the sample are restrained to allow for accurate and consistent results. The sample is drawn to its failure limits and the point at which the sample fails and tears apart is annotated. The results will vary from sample to sample because of changes in molecular composition, quality of materials used in manufacture, and quality of the manufacturing process; however, the sample must meet the minimums for acceptance in the building process. The strain at failure can be observed and measured. If a sample returns to its original shape or length after the test, it is said to be **elastic**. If the sample does not return to its original form then a breakdown of the material has occurred and permanent deformity will occur; this is referred to as *yielding*. The subsequent deformation causes a change in the properties of the sample known as *plastic deformation*. The sample can be classified as either ductile or brittle. A sample that can be subjected to large strains before it ruptures is called a **ductile** material (see Figure 1-7). Mild steel, brass, and zinc will exhibit the characteristics of being ductile.

A sample that exhibits little or no yielding before failure is referred to as **brittle** (see Figure 1-8). Often these samples do better in the compression phase of the test as cracks will close and the sample will bulge prior to failure. Cast iron and concrete are examples of brittle materials.

elastic ■ The ability of concrete, metal, or wood members to return to size and shape after deformation under load.

ductile ■ Capable of being hammered into thin layers or drawn out into wires; the ability of metals to be subjected to stress without breaking.

brittle ■ The term pertains to the ability of metals to retain shape under loads.

Wind Load

Wind exerts a variety of forces on a structure as it encounters and then attempts to pass over and/or through it. Wind exerts force across the total surface of the structure as it meets it. The structure will spread the force collectively across its surface, but it will also be affected by the same force proportionately as well as by the speed of the wind versus the construction materials and shape of the structure (see Figure 1-9). Simply put, the face

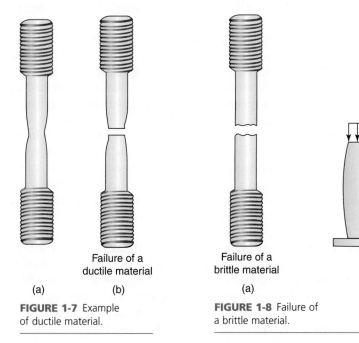

Failure of a
ductile material

(a)   (b)

FIGURE 1-7 Example
of ductile material.

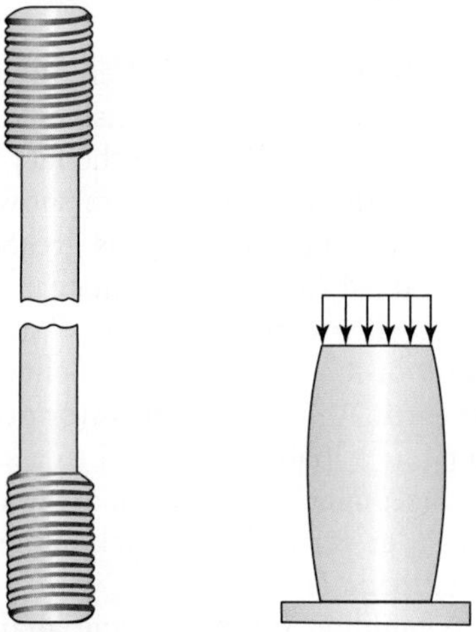

Failure of a
brittle material

(a)   (b)

FIGURE 1-8 Failure of
a brittle material.

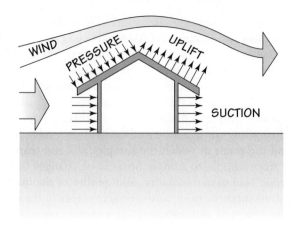

FIGURE 1-9 Wind loads on a structure. *Source: Onouye, Barry S.; Kane, Kevin, Statics and Strength of Materials in Architecture and Building Construction, 2nd Edition, © 2002, p. 8. Reprinted by permission of Pearson Education, Inc., Upper Saddle River, NJ.*

of the structure spreads the force of the wind along its surface area for as long as it can withstand the wind speed and subsequent exponential forces.

A masonry structure will withstand the effects of the wind better than one constructed of wood. A structure presenting a vertical flat surface will absorb more force than one constructed in a geodesic or angular shape. Wind conditions will be different between the ground level and multiple stories above. These forces seek to cause the building to move, to fail in a lateral way. In high-rise structures, sway factoring has to be engineered into the construction techniques so that occupants on the upper floors are not subjected to perceptible movement of the building as it flexes in response to the force of the wind.

Snow Load

snow loads ■ Mathematical calculations of the force imposed upon a roof by deposited snow. The wetness and depth of the snow are correlated relative to the pitch of the roof.

Snow loads are factored into the design and size of the roof structural members based on the anticipated worst possible scenario or the anticipated live load (for example, workers performing maintenance or equipment in place temporarily). A factor of 20 pounds per square foot (psf) is commonly used for calculating live loads for roof assemblies. Snowfall amounts for an area are established after historical data have been collected and evaluated (see Table 1-2).

Snow loads are calculated differently for flat roofs versus pitched roofs (those having slope). Snow load is generally considered to be the same for flat roofs as for snow that has fallen on the ground (despite the fact that the wind will most likely blow some or all of the snow off the flat roof before snow blows off the ground). For sloped or pitched roofs, however, if the roof angle is greater than 20 degrees or the ground snow load is greater than 20 psf the requirements can be reduced.

point loading ■ Imposing a load on a small area relative to its mass.

impact load ■ A load delivered unexpectedly to a specific point.

When a load is concentrated in a small area it is referred to as **point loading**. This concentration of load must still be delivered through various components until it safely reaches the earth. A large drill press (or other similarly heavy equipment) is an example of point loading. If a load is delivered to a small area in a sudden event, then it is called an **impact load** (see Figure 1-10). Impact loads can be delivered either horizontally, called a lateral impact, or vertically. A truck man stepping off a ladder onto a roof is an example of an impact load.

When a building starts to come apart because of failures caused by a fire or other type event, the loading of the various supporting members shifts. This transformation from a stable load platform to one that is dynamically changing is what often leads to early failures that can cause death or injury to firefighters. For example, assume a load is calculated for a floor assembly. The floor joists are factored in and the floor sheathing is acting to support stored materials. A fire starts below the assembly and significantly

TABLE 1-2 | Ground snow loads at selected locations

LOCATION	SNOW LOAD (PSF)	LOCATION	SNOW LOAD (PSF)
Huntsville, Alabama	5	Jackson, Mississippi	3
Flagstaff, Arizona	48	Kansas City, Missouri	18
Little Rock, Arkansas	6	Concord, New Hampshire	63
Mt. Shasta, California	62	Newark, New Jersey	15
Denver, Colorado	18	Albuquerque, New Mexico	4
Hartford, Connecticut	33	Buffalo, New York	39
Wilmington, Delaware	16	New York City, New York	15
Atlanta, Georgia	3	Charlotte, North Carolina	11
Boise, Idaho	9	Fargo, North Dakota	44
Chicago, Illinois	22	Columbus, Ohio	11
Indianapolis, Indiana	22	Oklahoma City, Oklahoma	8
Des Moines, Iowa	22	Portland, Oregon	8
Wichita, Kansas	14	Philadelphia, Pennsylvania	14
Jackson, Kentucky	18	Providence, Rhode Island	23
Shreveport, Louisiana	3	Memphis, Tennessee	6
Portland, Maine	60	Dallas, Texas	3
Baltimore, Maryland	22	Salt Lake City, Utah	11
Boston, Massachusetts	34	Seattle, Washington	18
Detroit, Michigan	18	Charleston, West Virginia	18
Minneapolis, Minnesota	51	Madison, Wisconsin	35

Source: Ellingwood, Bruce; Redfield, Robert (1983). "Ground Snow Loads for Structural Design," Journal of Structural Engineering. Used with permisison from ASCE.

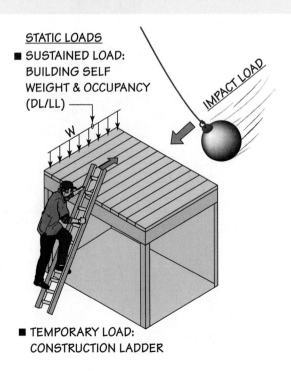

STATIC LOADS
■ SUSTAINED LOAD:
BUILDING SELF
WEIGHT & OCCUPANCY
(DL/LL)

IMPACT LOAD

■ TEMPORARY LOAD:
CONSTRUCTION LADDER

FIGURE 1-10 Loads based on time. *Source: Onouye, Barry S.; Kane, Kevin, Statics and Strength of Materials in Architecture and Building Construction, 2nd Edition, © 2002, p. 264. Reprinted by permission of Pearson Education, Inc., Upper Saddle River, NJ.*

burns the surface area of one or more joists. These fall away and now the stored materials are resting solely on the sheathing, which quickly fails, and in the subsequent collapse the loads begin shifting, causing other members to be overstressed and more collapses to occur. These collapses are caused by the sudden impact loading of an assembly.

This becomes very relevant when we consider trusses. A truss is a successful system when all members of the truss are sharing the load as a complete unit. If, however, the connections on a truss fail, or any of the independent units of the truss fail, the entire truss can and often does fail. In catastrophic outcomes other trusses collocated to the affected truss also fail.

Earthquakes, Hurricanes, and Tornadoes

Structures in earthquake-, hurricane-, and tornado-prone areas have to be constructed so that movement can be accommodated by the structure. For earthquake-prone areas, the foundation is the key. The foundation must be able to withstand lateral movement while supporting the loads above. The connections between the upper loads and the foundation must also be able to withstand movement (see Figures 1-11 and 1-12).

Foundations are often placed on **spreader footings,** which allow a contiguous sharing of the load and movement, while connectors keep the unit moving as one (see Figure 1-13).

The International Building Code (2009 edition) contains fourteen pages of formulas and specifications for handling the vibrations and harmonic motions associated with earthquakes. Earthquakes are caused when the tectonic plates shift at a fault line. Harmonic movement occurs as a result of the vibrations. The usual result of harmonic movement is that the building begins to shake as it attempts to keep up with the back and forth and undulation of the ground. For all of the calculations, specifications, and modern

spreader footings ■
A footing whose sides slope gradually outward from the foundation to the base.

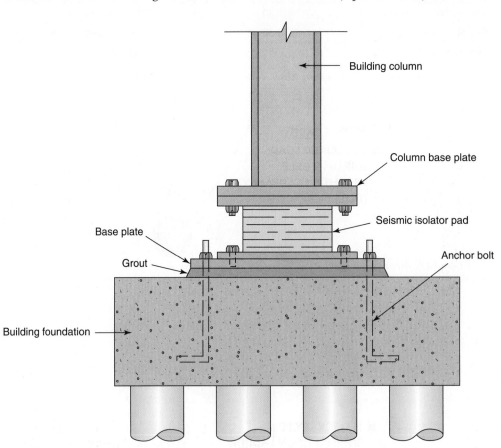

FIGURE 1-11 Seismic isolator pad for building column. *Source: Andres, Cameron K.; Smith, Ronald C., Principles and Practices of Commercial Construction, 7th Edition, © 2004, p. 127. Reprinted by permission of Pearson Education, Inc., Upper Saddle River, NJ.*

FIGURE 1-12 Seismic isolator pad for building column.

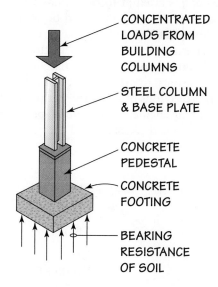

CONCENTRATED LOADS FROM BUILDING COLUMNS

STEEL COLUMN & BASE PLATE

CONCRETE PEDESTAL

CONCRETE FOOTING

BEARING RESISTANCE OF SOIL

FIGURE 1-13 Spreader footings. *Source: Onouye, Barry S.; Kane, Kevin,* Statics and Strength of Materials in Architecture and Building Construction, *2nd Edition, © 2002, p. 214. Reprinted by permission of Pearson Education, Inc., Upper Saddle River, NJ.*

methods being employed, it can be said that there is no such thing as an earthquake-proof building. California has, however, made significant progress towards designing and constructing buildings better able to sustain earthquakes. California has also significantly completed retrofitting buildings subject to earthquake damage. Los Angeles and San Francisco have developed model building codes for this purpose.

The key during earthquakes is to implement collapse zones, limit entry into zones for rescue only (not for searching), and be prepared for significant aftershocks that may be even more powerful than the initial jolt. Aftershocks can be defined as one or more smaller earthquakes occurring after a previous larger earthquake in the same area. If the aftershock is larger than the original earthquake, it then becomes the main shock and the original main shock is designated as the foreshock.

During hurricanes and tornadoes, the force of the wind causes overpressurization and vacuum effects simultaneously on the structure. Relief is gained by stronger connections between all main members and the foundations. **Torsion** loads must also be dealt with. This twisting action will first affect connections. When those fail, main components are affected and begin to twist as they fail (see Figure 1-14).

torsion ▪ The result of twisting.

Torsion can be offset by volume. A main component can be built at a size able to withstand the twisting. The effects of torsion can also be controlled by restraint. The

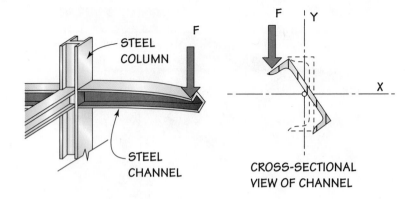

FIGURE 1-14 Torsion on a cantilever beam. *Source: Onouye, Barry S.; Kane, Kevin, Statics and Strength of Materials in Architecture and Building Construction, 2nd Edition, © 2002, p. 57. Reprinted by permission of Pearson Education, Inc., Upper Saddle River, NJ.*

STEEL COLUMN

STEEL CHANNEL

CROSS-SECTIONAL VIEW OF CHANNEL

buttresses ▪ Projecting structures built against a wall or building to provide greater strength and stability.

pilasters ▪ Rectangular columns attached to a wall or pier for stiffening. Structurally it's a pier but it is treated as a column with a capital, shaft, and base.

system employed to restrain the movement of a structural component, such as a column, can counteract the column's attempt to twist under load and fail. This can be accomplished by securely connecting both the top and bottom of the column and also by attaching a horizontal connector at the midpoint between two columns in the vertical axis of the columns. Another example is the use of **buttresses** in masonry systems (see Figure 1-15). The force of a gabled roof trying to flatten and the built-in weakness of masonry laterally cause the masonry to attempt to twist along its longitudinal axis. The use of buttresses, horizontal projections of masonry on the exterior of a structure extending from the ground vertically at engineered positions, prevents this condition. Interior applications of this system are called **pilasters** and they accomplish the same task (see Figure 1-16).

FIGURE 1-15 Forces in a buttress system. *Source: Onouye, Barry S.; Kane, Kevin, Statics and Strength of Materials in Architecture and Building Construction, 2nd Edition, © 2002, p. 25. Reprinted by permission of Pearson Education, Inc., Upper Saddle River, NJ.*

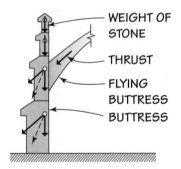

WEIGHT OF STONE

THRUST

FLYING BUTTRESS

BUTTRESS

FIGURE 1-16 Pilasters. *Source: Onouye, Barry S.; Kane, Kevin, Statics and Strength of Materials in Architecture and Building Construction, 2nd Edition, © 2002, p. 211. Reprinted by permission of Pearson Education, Inc., Upper Saddle River, NJ.*

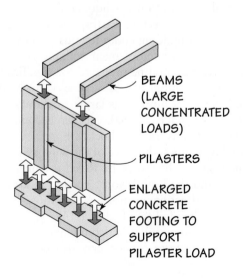

BEAMS (LARGE CONCENTRATED LOADS)

PILASTERS

ENLARGED CONCRETE FOOTING TO SUPPORT PILASTER LOAD

Forces

The best method for distributing loads is by compression. Trusses and concrete are fantastic under compression. All loads are delivered directly down through the material or system. (This can be demonstrated by pushing your fist down into your palm.) Conversely, trusses and concrete are completely inadequate against tension or the suspending of loads. (This is similar to doing chin-ups.) If you have ever demolished a concrete slab on the ground then you know that as long as the slab is in contact with the earth it is extremely difficult to break up, but if you raise the slab even a fraction of an inch, the slab breaks much more easily. This is switching the force on the slab from compression to tension. Trusses are engineered to share the load delivered through a system of geometric shapes held together by connectors (see Figure 1-17). In compression, the loads are designed to be delivered from above. If the connectors begin to fail, then the loading shifts from compression to tension and total failures will occur.

The collapse of the Hyatt Hotel skyway in Kansas City was caused when inferior connecting nuts were used in construction and their weight limitations were reached and then exceeded. The skyway was designed to be suspended in tension, and if the connections between the suspension cables and the platform had been according to specifications the system would have held. The after-action investigation revealed that nonrated hardware had been used and the weight of the people overtaxed the system, resulting in more than 115 deaths.

For those involved in acceptance inspections, or even those performing familiarization inspections, some knowledge regarding bolts is essential. Most bolts are made from low- or medium-grade carbon steel, but they can also be made from alloy steel for heavier-duty applications. We will study steel in a later chapter. A bolt's ability to withstand deformation and breakage is what qualifies it for a specific strength grade or property classification (see Table 1-3). If a bolt is destined to fulfill a heavy-duty rating, the bolt will undergo a process known as quenching and tempering, which involves heating the bolt to approximately 1000°F and then rapidly cooling it, followed by reheating the bolt and then cooling it gradually after its initial shaping.

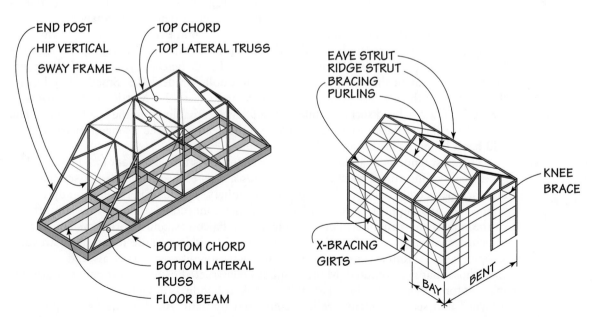

FIGURE 1-17 Typical use of trusses in bridges and buildings; note the geometric shapes. *Source: Onouye, Barry S.; Kane, Kevin,* Statics and Strength of Materials in Architecture and Building Construction, *2nd Edition, © 2002, p. 142. Reprinted by permission of Pearson Education, Inc., Upper Saddle River, NJ.*

TABLE 1-3	U.S. and Metric Grading of Bolts					
	HEAD MARKING	**GRADE AND MATERIAL**	**NOMINAL SIZE**	**PROOF LOAD**	**MINIMUM YIELD STRENGTH**	**MINIMUM TENSILE STRENGTH**
U.S. Standard		**Grade 2** Low or Medium Carbon Steel	1/4–¾ inches	55,000 psi	57,000 psi	74,000 psi
		Grade 5 Medium Carbon Steel, Quenched & Tempered	1/4–1 inches	85,000 psi	92,000 psi	120,000 psi
		Grade 8 Medium Carbon Alloy Steel, Quenched & Tempered	1/4–1½	120,000 psi	130,000 psi	150,000 psi
Metric	8.8	**Class 8.8** Medium Carbon Steel, Quenched & Tempered	0–16 mm	84,000 psi	93,000 psi	116,000 psi
	10.9	**Class 10.9** Alloy Steel, Quenched & Tempered	5–100 mm	120,000 psi	136,000 psi	151,000 psi
	12.9	**Class 12.9** Alloy Steel, Quenched & Tempered	1.6–100 mm	141,000 psi	160,000 psi	177,000 psi

Fine threads are stronger than coarse threads in tension and shear strength. This is due to the larger stress area (spread out over more threads) and the larger minor diameter (the diameter as measured from the inside of the threads) of a fine thread bolt.

On the chart you can see that there are three strength categories, representing stress levels, that the bolt must endure as part of a given grade or classification. *Proof load* is the axial tensile load the bolt must withstand without evidence of permanent deformation. *Yield strength* is the maximum load at which the bolt exhibits a specific permanent deformation. *Tensile strength* is the maximum load in tension (pulling apart) that the bolt can withstand before breaking or fracturing. From the chart you can see that there are visual cues as to the strength and values of a bolt. Grade 2 U.S. Standard and Class 8.8 Metric are the weakest while Grade 8 U.S. and Class 12.9 Metric are the strongest. Grade 2 has a flat-head surface while Grade 8 has 6 radial lines on the head surface. The Class 8.8 is the weakest metric and has 8.8 on the head while the strongest metric is 12.9 and has 12.9 on its head surface.

shear ■ Shear is caused by opposing forces that almost meet head on. This type of loading depends heavily on the connectors.

The most tenuous of all load delivery mechanisms is through **shear** (see Figure 1-18). (Shear can be illustrated by sliding your hands across each other.) These assemblies are attached perpendicular to their supporting members. The load is resting totally on the connectors and support of the load relies on the connectors' ability to remain affixed to the supporting structures. For example, decks, balconies, and flooring systems can be attached in shear. If a nonrated connector is used for this purpose it can fail with catastrophic results. Rated connectors are constructed of different materials and processes than those that are nonrated. Many buildings constructed of balloon framing utilize this method for attaching the floor assemblies. Look at buildings in your area that have decks attached and see how many are just connected by nails. All these decks are potential

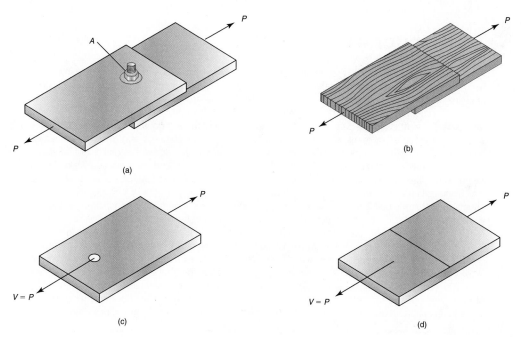

(a)

(b)

(c)

(d)

FIGURE 1-18 Examples of single shear (sliding) connections in wood and steel. *P* = total force; *V* = single shear force; *A* = moment where force occurs (connector).

disasters. A 16-penny (16p) cement-coated nail will support 160 pounds. The cement coating increases the nail's ability to adhere to the wooden members; as the nail is driven in, heat is created, causing the cement coating to temporarily liquefy and become an adhesive. But the movement of occupants either walking or dancing causes vibrations, which can withdraw the nail from its attachment. The correct method is to install a bolt completely through the header board into the main structure where it is connected to a washer and nut. This method is called *through bolting* (see Figure 1-19).

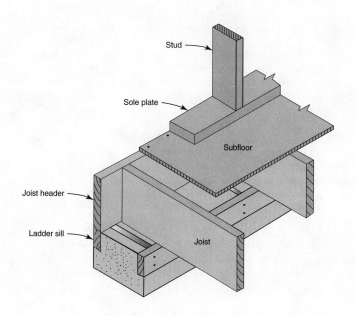

Stud

Sole plate

Subfloor

Joist header

Ladder sill

Joist

FIGURE 1-19 Through bolting must be installed at the connection of the joist header on the deck or addition, and into the main structure.

SCHOOL OF HARD KNOCKS

I was a lieutenant in command of an engine company during the mid-1980s. The engine company, Number 10, was the busiest company in the city. We responded to a three-story Type III constructed dwelling at approximately 2100 hours on a cold and windy evening. The structure measured 25' × 60'. Fire was extensive throughout the first floor and migrating to the second floor. I had fought many fires in this type of structure. We attacked the blaze with a 1½" stream and were making good progress when the entire second floor collapsed in on us, trapping the three members of my company and myself. Nobody was seriously injured, but that incident caused me to change the way I approached the combat of firefighting forever. After we had been extricated, which took about thirty minutes, and the remaining fire had been extinguished, I began to question the others as to whether they had witnessed any of the cues that indicate impending collapse. They answered that they had not noticed anything unusual. It was determined by the fire marshals that the building had been extensively remodeled. The floor joists were no longer embedded in the masonry walls, but had been reconfigured utilizing lightweight hangers attached to boards anchored to the brick face; the joists were not in compression as they should have been but rather were in shear.

I'm not completely blameless, however. When we first went into the building I should have recognized immediately that the stairs to the second floor were not in a normal position. In all the other dwellings in the neighborhood the stairs for the upper stories were just inside the front door with the basement stairs stacked directly underneath. These stairs were in an L-shape in the rear. I had remodeled more than fifty of these dwellings as a carpenter, so there was no excuse for my missing that key piece during my size-up. This was an important lesson for me on the forces of nature; luckily we didn't pay a price in human life.

FIRST RESPONSE

Buildings Under Construction/Demolition or Significant Renovation

Buildings under construction, those being demolished completely or those under significant renovation share some common problems for firefighters and other emergency personnel. Too many firefighters are injured and killed each year responding to these incidents. Many of the causes for these events are listed below:

- The structures are not complete.
- Many systems (such as sprinklers, fire pumps, or elevators) are not functional.
- Fire can access areas easily because of exposed spaces.
- The application of water by the use of hoselines can compromise structural integrity. (Attempts to save nineteenth-century facades for historical purposes will expose the mortar to degradation by fire department hose streams and cause collapse.)
- Egress and ingress points can be compromised, for example, when stairs are removed and replaced with scaffolding.

  Tactical considerations include the following:

- Respond to these structures with emphasis on firefighter safety.
- Firefighter accountability and the use of the incident command system is paramount.
- Never allow entry without charged hoselines readily available.
- Ensure adequate ladder placement for firefighter retrieval from upper floors.
- Enter the structure only if absolutely necessary.
- If significant fire conditions exist, respect collapse potential, particularly when operating around masonry.
- If concrete formwork is present assume the concrete is not fully cured and operate above or beneath the formwork.
- Be cognizant that openings intended for stairs or elevators may be uncovered or loosely covered by plywood, old doors, or even tarps.

Fire department and emergency medical services units are dispatched to a location upon a report of workers trapped at a residential construction site. Upon arrival, the units discover three workers pinned under debris. The debris consists of lightweight steel, wood, and gypsum board. The job foreperson tells the fire department personnel that workers had been relocating a steel beam in the basement when the beam slipped from the support column and the floor above failed.

1. What concerns would you have in regards to removing the debris?
2. What information do you need for calculating the loads being imposed on the workers?
3. How would you stabilize the remaining structural members?

Summary

Gravity is a natural force that acts on all buildings and their members at all times. Gravity wants to pull down the structure. Builders and engineers calculate the best method for attachment to deliver all loads back to the earth. The method can be compression, tension, or shear. Nature or fire can influence load delivery systems by removing members and thereby switching loads around. Sudden load delivery or impact loads can cause the failure of assemblies; however, no consideration is given during design or construction to the potential effects of fire and the corresponding shifting of loads. Testing is conducted to determine the rated strength under the projected loads applied for all major structural components.

Emergency response personnel must relate the expected load shifts and potential failure points in their initial size-up. Scene commanders must be regularly updated on conditions occurring in the interior of the structure being affected. The on-scene commander must continue ongoing risk assessment analysis throughout the period that personnel are inside a structure or in the potential failure zones on the exterior. The forces acting upon the main structural members will be accelerated by the thermal effects of fire and the subsequent effects of gravity will occur quickly and often without audible or visible signs or warnings.

Review Questions

1. When does gravity begin to affect the structure?
2. Describe the differences between axial and eccentric loading.
3. What does yielding mean and what effect does it have on the sample?
4. Describe compression and give examples.
5. Describe tension and give examples.
6. Discuss and define shear as it relates to public and firefighter safety.
7. Discuss the difference between ductile and brittle and give examples.
8. How can the effects of torsion be overcome?

Suggested Reading

ANSI/ASCE 7-95. 1976. *Minimum Design Loads for Buildings and Other Structures*. Reston, VA: American Society of Civil Engineers.

Boston Fire Department http://www.cityofboston.gov/fire/memorial/vendome_fire.asp./

International Building Code. 2009 edition. International Code Committee. Washington, D.C.

NFPA 5000. 2003. *Building Construction and Safety Code* Quincy, MA: National Fire Protection Association.

Onouye, B. 2002. *Statics and Strength of Materials for Architecture and Building Construction*. Upper Saddle River, NJ: Pearson Education.

Smith, J. 2008. *Strategic and Tactical Considerations on the Fireground*. Upper Saddle River, NJ: Pearson Education.

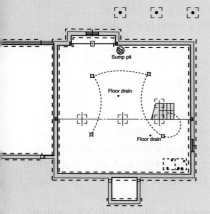

OBJECTIVES

After reading this chapter, you should be able to:

■ Understand the terms used in basic blueprint reading.
■ Recognize the different parts of a plan.
■ Define key terms contained in all plans.
■ Understand how codes are developed and the political processes involved.

Introduction

Any book concerning building construction would be incomplete without looking at codes and plans. Building codes delineate the level of risk that a community will be exposed to in regard to building collapse and failure. A code can be structured around many different standards. It can be specific or generalized. The code process is a political one whereby not only the end user is represented but also the banks and contractors who have a monetary interest in the process.

Plans are the road maps of the building process. These can be quite basic for a simple residential structure or quite complex for a large commercial building or a high-rise. A fireground commander must be conversant with plans and their symbols to better interact with the building management and other represented agencies during emergencies. When responding to incidents involving large buildings or high-rise structures, it is imperative that the fireground commanders understand building systems, their capabilities, and their limitations. This is necessary in order to be able to utilize building plans to communicate with the building staff, execute procedures, or react to the building's limitations. For example, not all HVAC (heating, ventilation, and air conditioning) systems can be manually operated to ventilate smoke and heat from specific floors.

Building Codes

Hammurabi, the king of Babylon (1796 BC–1750 BC), instituted the first building code in 1790 BC as part of his Code of Hammurabi. One of the most striking of its requirements called for the execution of the builder and architect of any structure that collapsed

FIREHOUSE DISCUSSION

On February 23, 1991, the Philadelphia Fire Department (PFD) fought one of its toughest fires, which occurred in the One Meridian Plaza building. One Meridian Plaza, originally known as the Fidelity Mutual Life Building, was a thirty-eight-story high-rise office building, 243 feet long by 92 feet wide, constructed between 1968 and 1972. At the time, it had been the tallest building built in Philadelphia since the 1930s. The building was constructed of internal skeletal-steel members, which were encased within sprayed-on fireproofing. The floors were constructed of concrete-on-metal pan floors. The exterior was granite curtain walls. There were three stairwells; wet standpipes were in two of them.

The PFD lost three of its members and another twenty-four firefighters were injured. The PFD struck twelve alarms, which brought fifty-one engines, fifteen ladder companies or trucks, eleven specialized units, and over 300 firefighters. The fire lasted nineteen hours.

At the time of construction, the city of Philadelphia only required sprinklers in service levels below grade in high-rise buildings. In 1984, the city adopted codes that required sprinklers in all new buildings, but only required them in existing buildings if significant renovations occurred to any floor and then only required the sprinklers for that floor or floors. At the time of the fire, only four floors had been entirely sprinklered with three more having partial protection. There was a supervised alarm system in place with two guards on duty and a building engineer.

The fire began on the twenty-second floor when rags soaked in linseed oil, used for refinishing woodwork, spontaneously combusted and set fire to other combustible solvents being stored on the work floor. Smoke detectors were not installed on the entire fire floor, which led to a delayed alarm. When the alarm sounded at 2023 hours, the engineer went up to the twenty-second floor where he encountered fire and smoke conditions. He radioed down to the guard at the front desk who subsequently called the alarm company to verify that the alarm was genuine. But the fire department was never called. The alarm was called in by a passerby, which led to a delay of many precious minutes. The PFD arrived on the scene at 2027 hours and began ascending the stairwells to battle the blaze. When they reached the eleventh floor, the building lost all electrical power due to heat entering an unprotected enclosure disabling electrical cables. The primary and backup electrical systems were in the same raceway enclosure. The natural gas-powered emergency generator also failed. The power could not be restored so the elevators and all electric systems were out of service.

On the fire floor the companies were unable to produce effective attack streams. The hoselines needed 100 psi to be effective, but were only producing 32 psi due to improperly adjusted pressure reducing valves (PRVs). These valves are installed so that pressure required to produce effective streams on the top floors do not overpressure the lower floors. In the engine pressure formula for hoselines, this is accounted for by the instruction to "add 5 psi per floor to accommodate lift."

The fire department battled the problems and the fire for over eleven hours. During this time, an engine company with three members was ordered to ascend the stairwell to the roof to open a door for ventilation. They became disoriented, ran out of air, and succumbed. Concerns arose that the building was becoming structurally untenable and all firefighters were ordered out of the building at 0700 hours. The fire continued to burn horizontally and vertically until it reached the thirtieth floor, which was the first completely sprinklered floor. Ten sprinkler heads activated and contained the fire until it ran out of fuel.

THINK ABOUT IT!

- What does your code require for sprinklers in high-rise buildings?
- Determine if your town or city uses a performance code or a specification code, or does it have a hybrid of both?
- Try to interview either a building official or an inspector to see how the code acceptance process works in your town or city.
- How much pump pressure would have been required to produce effective streams for six 1¾" handlines for two floors of fire assuming over a 50 percent involvement on floors twenty-two through twenty-six?
- Without elevators, how would you get all of the equipment and personnel to the fire floors, and how long do you estimate that it would take?
- How many personnel do you have available for a fire of this magnitude?
- Discuss the possible alternative plans for fighting a fire of this magnitude.
- Read about some other high-rise fires in this country and determine if there are any similarities to the fire described above.

and killed its occupants. In the seventeenth century, after many catastrophic fires had occurred, London forbade the use of a thatch or straw roof. Also in the seventeenth century, the city of Boston forbade any homeowner from constructing a chimney of wood. Unfortunately, many of our codes today have evolved from disasters.

The first building code in the United States was developed in 1905 by the Fire Underwriters. Building codes are legal documents. They are enforced through the power of the states and delegated to local municipalities. State building officials supervise or are responsible for building officials at the city, county, and town levels.

A building code is a collection of laws listed in booklet form that apply to a community that adopts its provisions. The code covers all aspects of the construction of a new building as well as the restoration or remodeling of an existing structure. The code also deals with the demolition of or repairs to existing buildings. The basic purpose of codes is to define the minimum sizes of materials that can be used while preventing collapse and that the basic protection of the inhabitants can be assured within reason. Codes will deal with standards of performance and material specifications as well as minimum requirements concerning such design factors as room size, ceiling heights, lighting, and ventilation.

CODE DEVELOPMENT

Under the Constitution, the right to regulate building construction rests with each state; however, prior to 1960 most state entities had delegated that right to individual cities and towns. Of course this led to a convoluted, nonstandardized system to which the construction industry and all its subsidiary manufacturers and suppliers had to respond safely and economically.

specification code ■
A code that requires exact specified materials and/or processes to accomplish a task or requirement of building.

performance code ■
A code that does not specifically define how an assembly is to be fabricated or a specific material is to be used.

There are two types of codes. A **specification code** establishes building construction requirements by reference to particular materials and methods. For example, a wall must be constructed on the exterior of a residential structure using nominal 2 × 4s spaced 16″ on center (o.c.) with three 16p nails at each connection. This type of code limits the construction process to exact types of materials and methods.

The second type of code is the **performance code**. It does not limit the selection of materials and systems but rather sets a requirement for performance. For example, a set requirement to withstand a stated load can be met if an engineer has tested the system proposed and certifies that it will accomplish the goal. For many states this is how truss systems came into wide use, because they were not recognized under specification codes.

As you read a code, you will see many references to standards or certain tests. Specifications become standards when they are adopted for use by a broad group of manufacturers, users, or specifiers. It is important to note that the development of standards included within a code involves many players, as described next.

National Fire Protection Association (NFPA)

The NFPA was organized in 1896 to promote the science and advance the methods of fire protection. It is a nonprofit educational organization that publishes and distributes various publications on fire safety. These publications include model codes, materials standards, and recommended practices. The NFPA maintains standing committees, referred to as technical committees, which prepare the standards for adoption at the annual meeting of the NFPA. The technical information is contained in a ten-volume compilation known as the National Fire Codes. The National Electrical Code is part of Volume 5. The disclaimer on the inside flyleaf of the NFPA 5000 states that the code was developed by consensus by a group of volunteers representing many interests and that this process has been approved by the American National Standards Institute (ANSI). Further, the NFPA disclaims liability for any entity that adopts the standards.

American National Standards Institute (ANSI)

ANSI was founded after World War I to standardize the production of war materials. It was initially known as the American Standards Association (ASA). The ASA acronym might be recognizable to photographers who use ASA film speed designations. The name was changed to the United States of America Standards Institute for a short time and became ANSI in 1969. ANSI does not develop its own standards but acts as a clearinghouse for voluntary standards developed by other organizations in the United States. It provides a nationally recognized status to standards set by private organizations.

American Society for Testing and Materials (ASTM)

ASTM was founded in 1898 as the American Section of the International Society for Testing Materials and was incorporated in 1902 as the American Society for Testing Materials. The current name was adopted in 1961. The ASTM is a nonprofit, privately funded organization. It brings together committees (over 2080) and subcommittees (135) comprised of representatives from producers and product users as well as members of the architectural and engineering communities.

ASTM standards are voluntary consensus standards. These standards are not mandatory unless required by a code or standard that references them. This probably sounds confusing and it is. Simply put, if a code or standard references an ASTM standard for adoption then for that code, the standard has been adopted by that jurisdiction. The word *consensus* implies that the members of the committee agreed on its final wording.

ASTM standards have a unique alphanumeric designation, consisting of an uppercase letter (alpha) followed by a number (of one to four digits), a hyphen, and finally the year of issue. A typical standard is designated as ASTM E 84-01, *Standard Test Method for Surface Burning Characteristics of Building Materials*. The alpha letter refers to a general classification of the test procedure or material (see Table 2-1).

ASTM standards can be revised by a committee at any time but must be reviewed every five years. Thus the year designation at the end may contain a lowercase alpha letter. For example, the standard could read ASTM E 84-01a. The *a* indicates that a revision was made. A *b* would indicate two revisions, *c* three, and so on. If a standard was approved but unchanged then the year of approval appears in parentheses as (2005).

Underwriters Laboratories (UL)

UL was established in 1894 by the insurance industry to counteract the huge number of claims being paid due to problems created by the new electrical and mechanical devices being used by the building industry. The UL tests approximately 80,000 products annually. The UL label indicates that the tested product is free from any danger. The Underwriters Laboratories is a not-for-profit organization and receives its financial support through testing of materials and the sale of publications.

The UL has approximately 560 safety standards, most of which are approved as ANSI standards. The UL is active in standard-writing committees for ANSI, ASTM, and the NFPA.

Trade Organizations

Trade organizations exist to promote and protect the interests of their members. The American Wood Preservers Association (AWPA) represents the manufacturers who treat wood products with fire-retardant chemicals and/or preservatives. The AWPA also provides a nationally accepted grading stamp system to designate levels of quality. For example, each piece of lumber or sheet of plywood carries an AWPA stamp.

Until 2000 there was not a national code enforced throughout the United States. Instead there were three *model codes* The Uniform Building Code was primarily used in the midwest and the western United States. The Standard Building Code was followed in

TABLE 2-1	ASTM Designation System
LETTER DESIGNATION	**TYPE OF MATERIAL OR TEST**
A	Ferrous Metals
B	Nonferrous Metals
C	Cementatious, Ceramic, Concrete, and Masonry Materials
D	Miscellaneous Materials
E	Miscellaneous Subjects
F	Materials of Specific Applications
G	Corrosion, Deterioration, and Degradation of Materials
ES	Emergency Standards
P	Proposals

TABLE 2-2	Model Codes	
NAME OF BUILDING CODE	CODE ORGANIZATION	ORGANIZATION HEADQUARTERS
Uniform Building Code (UBC)	International Conference of Building Officials (ICBO)	Whittier, California
Standard Building Code (SBC)	Southern Building Code Congress International (SBCCI)	Birmingham, Alabama
National Building Code (NBC)	Building Officials and Code Administrators International (BOCA)	Country Club Hills, Illinois

the southeastern United States, particularly along the Gulf of Mexico and Atlantic Ocean coasts. The National Building Code was used in the northeastern sections of the United States. The three codes were created in an effort to begin standardizing information that went into the development of the *CABO One- and Two-Family Dwelling Code* in 1971. This code was for all detached dwellings less than three stories in height.

These three codes, referred to as the legacy codes, were considered to be the state of the art. In 1994, the three model code groups convened to combine their efforts to create a singular code without regional limitations. They formed the International Code Council (ICC) to accomplish this task (see Table 2-2).

International Building Code (IBC) and International Residential Code for One- and Two-Family Dwellings (CABO)

For many years, the staffs working on the three main building codes studied ways in which common ground could be reached so that one national code could be made applicable in all regions of the country. In February 1998, the first draft was distributed and by 1999 the principals of all three codes approved the second draft of the proposed national code. The new International Building Code (IBC) came out in 2000 and is designed to replace the other three existing codes. The IBC is updated every three years. The last update was published in 2009. The referenced standards are also published and updated every three years.

The new code differs in that it is almost exclusively a performance-based code. It also contains new language and terminology. It places emphasis on the metric system, whereas the previous code was restricted to the English system. It contains changes to occupancy classifications and types of construction, and is a valuable step towards more standardization in this country regarding the construction process. Any city or county government can, however, write its own code or continue to adopt the previous local national code. The acceptance of this new code is not mandatory, so it is best to be diligent about updating your knowledge of your own local codes.

This is especially important in areas where there isn't a designated building official or where use and occupancy permits are not required prior to habitation. The role of the building official is to act as the senior individual responsible for code development, acceptance, and enforcement. This individual usually has vast experience in the construction field and can adequately provide trained inspectors, settle code disputes, and ensure that all citizens within the locale will be guaranteed that at least the minimums for safety have been complied with. Use and occupancy permits ensure that all phases of the construction process have been inspected, all code items have been enforced, and workmanship has complied with good standard practices. It is, however, up to individual states to choose which code or parts of codes they will adopt or, as in the case of New York City and Chicago, allow some cities to maintain their own codes independent of any other entities.

The International Residential Code for one- and two-family dwellings has replaced the existing CABO one- and two-family code. The new code addresses the latest technological advances in building design and construction. Among the changes are provisions for steel framing and energy-saving initiatives. It also contains provisions relative to mechanical, fuel gas, and plumbing that coordinate with the International Mechanical Code and International Plumbing Code. This code is also more performance-based than the previous code, which was a specification-based code.

OCCUPANCY CLASSIFICATIONS

Codes in general are designed to reduce the dangers faced by occupants. These can include calculating how fire will spread based the structure itself, identifying the presence of toxic and heated gases, unprotected openings in the structure, combustible decorations or wall finishes, and finally the number of occupants allowed within the structure at one time. For this reason, many of the codes group the occupancies relative to the probability of certain numbers of occupants. The more people and/or hazards present in the structure at one time, the more stringent the requirements will be for the protection of the structure and any internal assemblies or decorations (for example, trim) against the spread of fire or its by-products.

The 2009 IBC consists of ten major occupancy groups with thirty-two divisions or subparts. These are:

A   Assembly
B   Business
E   Educational
F   Factory and Industrial
H   Hazardous
I   Institutional
M   Mercantile
R   Residential
S   Storage
U   Utility

For the purposes of illustrating the differences between the requirements, we will look at two of these occupancy groups as they translate to structural requirements, fire and smoke control, and other safety factors. *Note*: A—Assembly and R—Residential.

ASSEMBLY

This group is subdivided into five divisions.

A-1:   A building or portion of a building having an assembly room with an occupant load of 1000 or more and a legitimate stage.

A-2:   A building or portion of a building having an assembly room with an occupant load of less than 1000 and a legitimate stage.

A-2.1: A building or portion of a building having an assembly room with an occupant load of 300 or more without a legitimate stage, including such buildings used for educational purposes and not classified as Group E or B occupancy.

A-3:   Any building or portion of a building with an occupant load of less than 300 without a legitimate stage, including such buildings used for educational purposes and classified as Group E or B occupancy.

A-4:   Stadiums, reviewing stands, and amusement park structures not included within other Group A occupancies.

RESIDENTIAL

The following are subdivisions of this group:

R-1: Hotels, apartment houses, and congregate residences (each accommodating more than ten persons).

R-3: Dwellings, lodging houses, and congregate residences (each accommodating ten or fewer persons).

The first difference between A and R is the number of occupants. The next difference is that A-1 can only be constructed of Type I assemblies while R-3 can be constructed of any of the types (I–V). However, R-3 is limited to three stories while A-1 is not. A major difference occurs in flame spread allowances. A-1 can have only a 0–25 rating while R-3 can have an index of 76–200. Many towns and counties interpret these occupancies to fit their own needs or political concerns. Therefore it is mandatory that you understand your particular code and how the occupancies are grouped.

GOING GREEN

Many cities and states are enacting changes to their codes to allow for more use of green assemblies or products, which utilize materials or methods to lower the carbon footprint of a structure, increase the use of renewable materials, lower the energy usage, assist in reducing rain runoff, and reduce the water usage by the inhabitants. This may pose some problems for firefighters, however. Some points to consider are:

- Green or brown roofs will reduce the ability to perform vertical ventilation (this will be discussed in depth in Chapter 4).
- Green roofs are, in fact, holding ponds, which will increase the risk of early collapse of a structure.
- The use of photovoltaic panels poses some safety hazards to firefighters (this will be discussed in Chapter 5).
- The use of advanced framing changes the structural load sharing (this will be discussed in Chapter 4).
- The approved use of structural insulated panels, SIPS, for replacement of wall and roof components will change load-sharing capabilities (this will be discussed in Chapter 4).
- The approval to replace cement masonry units with recycled plastic blocks will increase fire loading and reduce structural stability during a fire (this will be discussed in Chapter 4).

Fire Resistance and Flame Spread

conduction ■ Transfer of heat by direct contact.

convection ■ Heat transmission by natural or forced circulation or motion.

radiation ■ Transfer of heat through space by wave motion or rays.

Fire is defined as the rapid oxidation of a material in the chemical process of combustion, releasing heat, light, and various reaction products (smoke, for example). Fire and its subsequent by-products communicate or spread by three principal methods. These are conduction, convection, and radiation. **Conduction** is the transfer of energy through matter from particle to particle; in simpler terms, by contact or touch. Imagine a room: the fire begins in a chair that touches the floor and a wall, both of which also become involved when they reach their combustible temperature. **Convection** is the transfer of heat energy in a gas or liquid by movement of currents. The fire in our imagined room has grown so hot that the unburned gases and heat energy are now rising, then moving horizontally across the ceiling causing the window curtains to begin to burn at the top of the window. **Radiation** is electromagnetic waves that transport energy directly through space. The room across the hall from our imaginary room suddenly erupts into flame due to heat radiating from the fire room.

Fire has to be considered a living entity. It eats, grows, and thrives as long as there is sufficient fuel, heat, and oxygen. Fire codes are designed to thwart fire by limiting its ability to grow. Conduction can be stopped by insulation, convection can be stopped by a physical barrier, and radiation can be stopped by distance and shielding. Increasing or decreasing the space between two buildings or homes will increase or decrease the ability of fire to spread between the two.

Most victims of fire are not actually killed by the fire but by its by-products: smoke and hot gases. How quickly a fire moves through a structure—*flame spread*—is very important. The fire and/or its by-products can quickly overcome the trapped occupants. This was found to be true in the 1942 Cocoanut Grove Lounge in Boston, Massachusetts, and more recently in 2003 at the Station (bar or lounge) in Warwick, Rhode Island. In both cases the fire raced along combustible finishes within the structure, preventing the occupants from escaping.

To prevent such occurrences in A—Assembly occupancies, the use of combustible finishes or decorations is prohibited. Additionally, the structural components and their subassemblies must not contribute to the development of smoke or allow fire to spread beyond the confines of the initial area. This is called *fire resistance*.

Flame spread is measured by how fast an applied flame moves along a 19.5 foot tunnel. The time it takes for flame to spread over the tested material is gauged relative to flame spread in red oak. Red oak is used because it produces uniform results. It has an arbitrary rating of 100 while cement asbestos board has a rating of zero. Flame takes approximately 5.5 minutes to cover 19.5 feet of red oak. This process of testing is monitored and listed under the following standards, which are also ANSI standards.

- NFPA 251, *Standard Methods of Fire Tests of Building Construction and Materials,* National Fire Protection Association
- UL-263, *Fire Tests of Building Construction and Materials,* Underwriters Laboratories, Inc.
- ASTM E119, *Methods of Fire Tests of Building Construction and Materials,* The American Society for Testing and Materials

The results of the tests are evaluated as follows:

- If the time is less than 5.5 minutes, the formula used is 100×5.5 divided by 19.5.
- If the time is more than 5.5 minutes but less than 10 minutes, then the formula is 100×5.5 divided by the time it took for the flame to travel 19.5 feet plus half the difference between the result and 100.
- If it takes more than 10 minutes, then the formula is $100 \times$ the distance in feet traveled divided by 19.5.

This might appear to be confusing, and it can be, but it has been used since 1917 with little or no change.

The same test is used when testing for smoke generation; this test is referred to as E84. Photoelectric cells are employed to measure obscuration. The same constants are used, with red oak being rated at 100 and cement asbestos board at 0. The complete test is explained in the *NFPA Handbook*.

In NFPA 101, Life Safety Code, surface materials are classified as follows:

- Class A—0–25
- Class B—26–75
- Class C—76–200

In fire service terms, a conundrum exists: Materials can exhibit low flame spread but high smoke generation and still be allowed in assemblies because of the limitations of the testing process itself.

In addition to limiting flame spread and smoke generation, we must also be concerned with fire travel between compartments or floors. The ASTM E-119 test is conducted similarly to the other tests except that a gas-fed flame is introduced beneath the material tested. Thermocouples are used to measure the temperature on the opposite side of the material. If the temperature does not exceed 250°F, then the material is rated as fire resistive. If the material being tested is a structural member, it is loaded during the test. If at any time the temperature exceeds 250°F or the flame penetrates through the material tested, that time becomes the rating number. Door and window assemblies are tested as a complete unit. If a door fails at fifty-nine minutes, it would receive a rating of three-quarters of an hour, not one hour. With a four-hour rated column or wall, the fire or heat is supposed to be able to penetrate that door or wall after no less than four hours. Remember, the number is 250°F for pass/fail. Do you ever encounter temperatures greater than that on the scene?

The complete test process can be found in the *NFPA Handbook*. It must be noted at this time that no building is completely safe from fire or collapse. If enough heat is applied for enough time, all buildings will fail. Firefighters, and especially their commanders, must keep this in mind at all times.

TOMBSTONE CODE ENACTMENT

Too often it takes a major disaster to force politicians to enact codes or to enforce existing codes. The fire department is not usually at the table when codes are being discussed. The fire department only gets to inspect the portions of a code that has been adopted by the AHJ (authority having jurisdiction). The following are some examples of incidents that created tombstone code enactments.

Iroquois Theater Fire, Chicago, 12/30/1903 (602 dead)

- Exit doors must open outward.
- Exits must be clearly marked.
- Fire curtains must be made of steel.
- Theater management must conduct fire drills with ushers and theater personnel.

Triangle Shirtwaist Fire, New York, 3/25/1911 (146 dead)

- Exit doors cannot be locked.
- Flammable remnants must be collected.

Cocoanut Grove Nightclub Fire, Boston, 11/28/1942 (492 dead)

- Flammable decorations banned.
- Exit doors must swing outwards.
- Exit signs must be visible at all times.
- Revolving doors used as egress must either be flanked by a normal out-swing door or retrofitted so that they fold flat when pushed against.

Winecoff Hotel, Atlanta, 12/7/1946 (119 dead)

- National safety standards established and strictly enforced.

Our Lady of the Angels School, Chicago, 12/1/1958 (95 dead)

- Sweeping changes in school fire safety regulations enacted nationwide.

Studying the building codes in your area and developing a close relationship with the fire inspectors in your department will help you be a better fireground commander.

Plans and Blueprints

As mentioned earlier in this chapter, building plans and blueprints are the road maps for the erection of a structure. They also show the locations of key elements and the building's systems. Any officer who is an incident commander, or who wants to be, needs to be competent at reading plans and prints. Drawings are the original plans while blueprints are the copies or *prints*.

Construction drawings are based on general principles. The shape of a structure is described in orthographic (multiview) drawings made to scale. Figured dimensions describe the structure's size and are indicated by dimension lines, arrowheads, and extension lines. The following lines are used in all drawings:

- Centerlines are composed of long and short dashes, alternately and evenly spaced with a long dash at each end; at intersections the short dashes cross. Very short centerlines may be broken if there is no confusion with other lines.
- Dimension lines terminate in arrowheads at each end. On construction drawings they are unbroken. On production drawings they are broken only where space is required to insert dimensions.
- Leader lines are used to indicate a part or section to which a note or other reference applies. They terminate in an arrowhead or a dot. Arrowheads should always terminate at a line; dots are used within the outline of an object. Leaders should terminate at a suitable portion of a note, reference, or dimension.
- Break lines are used for brevity. For example, rather than continue a drawing of a lengthy, unbroken masonry wall, a break line will be drawn to indicate the continuation. Short breaks are indicated by solid freehand lines. Full, ruled lines with freehand zigzags are used for long breaks.
- Sectioning lines indicate the exposed surfaces of an object in a sectional view. They are generally full, thin lines, but they may vary with the kind of material shown.
- Extension lines indicate the extent of a dimension; they should not touch the outline.
- Hidden lines consist of short dashes evenly spaced and are used to show the hidden features of a part or assembly. They always begin with a dash in contact with the line where they start, except when such a dash would form the continuation of a full line. Dashes touch at corners. Arcs start with dashes at the tangent points (where they touch each other).
- Outline or visible lines represent those outlines of an object or assembly that can actually be seen.
- Cutting-place lines show where a section has been taken from the drawings to be shown again in greater detail.

Overall relationships of assemblies or parts are shown in general drawings similar to assembly drawings. Important specific features are shown in detail drawings, which are usually drawn to a larger scale than the general drawings. For example, the general plan might be scaled to 1/4" meaning that 1/4" is equal to 1'. The details might be scaled such that 1" is equal to 1', which provides better clarity. Additional information about size and material is furnished in the general and specific notes and will be indicated by a reference number.

TYPES OF DRAWING PLANS

Construction drawings present as much construction information as possible graphically (by means of pictures). A **plan view** is a view of an object as it would appear if projected onto a horizontal plane passed through or held above the object or assembly. The most common construction plans are **plot plans** (also called **site plans**), foundation plans, floor

plan view ■ In architecture, a diagram showing a horizontal view of a building taken from above.

plot plans ■ Plans showing the size of the lot on which the building is to be erected as well as all data necessary before excavation for the foundation is begun.

site plans ■ Similar to plot plans, they show the location of the building on the lot.

utilities ■ In building, the term usually applied to electricity, HVAC, and plumbing.

foundation plan ■ An architectural drawing showing the perimeter of building. The drawing includes the position of main girders, structural columns requiring footings, and door and window locations.

plans, and framing plans. Other plan types are architectural and engineering. A site or plot plan shows the contours, boundaries, roads, **utilities**, trees, structures, and any other significant features pertaining to or located on a site. The locations of proposed structures are indicated by appropriate outlines or floor plans. By locating the corners of a proposed structure at given distances from a reference or baseline, which is shown on the plan and can be found at the site, the plot plan provides essential data for those who will lay out the building lines. By indicating the elevation of existing and proposed earth surfaces, the plot plan guides graders and excavators. After the building has been completed the reference points are maintained for any possible future surveying concerns.

A **foundation plan** (see Figure 2-1) is a plan view of the structure projected on an imaginary horizontal plane passing through at the level of the top of the foundation. The plan shows the outlines of footings, the materials used to construct the foundation, and footings for piers or columns. Its details show rebar placement in the footings and piers. The notes indicate concrete mix ratios and slump values. If concrete masonry units (CMUs) are used the plan will call for their sizes and will indicate any areas requiring additional supports for weight distribution.

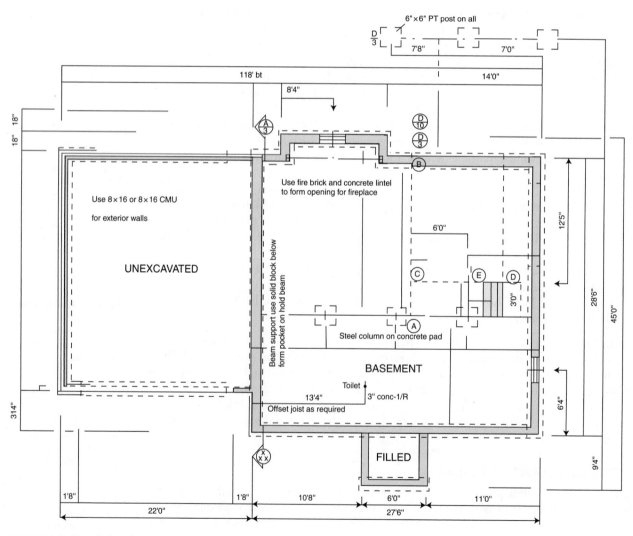

FIGURE 2-1 Foundation plan.

A **floor plan**, or building plan, shows the lengths, thicknesses, and composition of the building walls on any given floor (see Figure 2-2). It also shows the widths and locations of door and window openings; the position of partition walls; the number of rooms or offices; and the types and locations of utilities within a structure, such as HVAC, plumbing, electrical, and fire. If these are too complicated, or will make the drawings confusing, they are given their own pages.

Framing plans (see Figure 2-3) show the dimension, numbers, and arrangement of structural members in wood-frame construction. When steel is used there are often reference numbers indicating specific elements such as columns, girders, or beams. This is done because weight-bearing points are different throughout an assembly. If the building is concrete, then the details, specification section, and the notes are absolutely paramount. Among other details are the placement of wiring chases, piping, and conduit routing inside the concrete. The configuration of rebar is shown. *Wall framing plans* show information regarding studs, corner posts, bracing, sills, and plates. They give information regarding headers, lintels, and other horizontal members in the walls in the details. A *roof framing*

floor plan ▪ An architectural drawing showing the length and breadth of a building and the location of rooms within the building.

framing plans ▪ Architectural drawings showing the skeleton of a building. The roof, wall, and floor parts are shown in their correct configuration. Window and door dimensions are also listed.

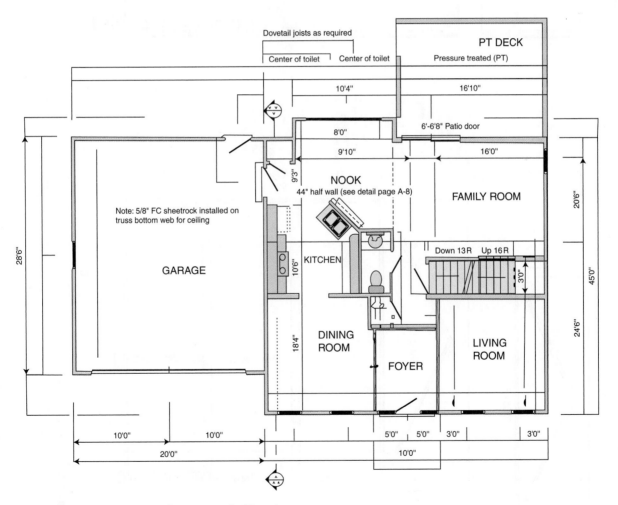

FIGURE 2-2 A floor plan, also referred to as a building plan.

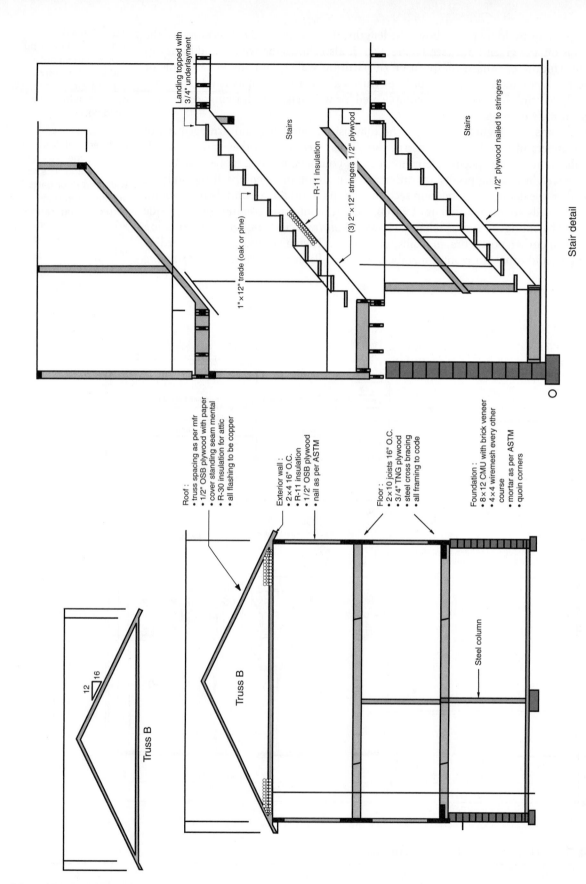

Roof :
• truss spacing as per mfr
• 1/2" OSB plywood with paper
• cover standing seam mental
• R-30 insulation for attic
• all flashing to be copper

Exterior wall :
• 2×4 16" O.C.
• R-11 insulation
• 1/2" OSB plywood
• nail as per ASTM

Floor :
• 2×10 joists 16" O.C.
• 3/4" TNG plywood
• steel cross bracing
• all framing to code

Foundation :
• 8×12 CMU with brick veneer
• 4×4 wiremesh every other course
• mortar as per ASTM
• quoin corners

Landing topped with 3/4" underlayment

Stairs

R-11 insulation

1" × 12" trade (oak or pine)

(3) 2" × 12" stringers 1/2" plywood

Stairs

1/2" plywood nailed to stringers

Stair detail

Truss B

12 ⎿16

Truss B

Steel column

FIGURE 2-3 Framing plan.

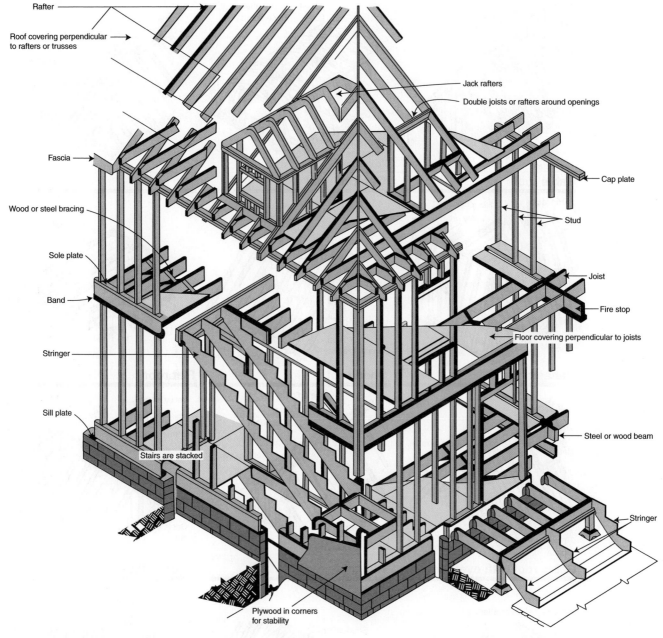

FIGURE 2-4 Size-up requires knowledge of techniques and terms.

plan shows information regarding rafters, purlins, ridges, truss placement, positions of penthouses or dormers, and any other appurtenance (see Figures 2-4 and 2-5).

A *utility plan,* also called a **mechanical**, shows the layout of heating, electrical, plumbing, or other system plans and their locations (see Figures 2-6, 2-7, and 2-8). Sizes of ducts, pipes, conduits, and locations of valves and fittings are some of the information shown on these plans. The plumbing plans show all pumps, fixtures, and valves necessary to install the system. A solid knowledge of plumbing symbols is necessary to read

mechanical ■ Drawing that specifies the electrical, HVAC, and plumbing systems in the building. It includes pipe dimensions, location of units, and all other specifications required.

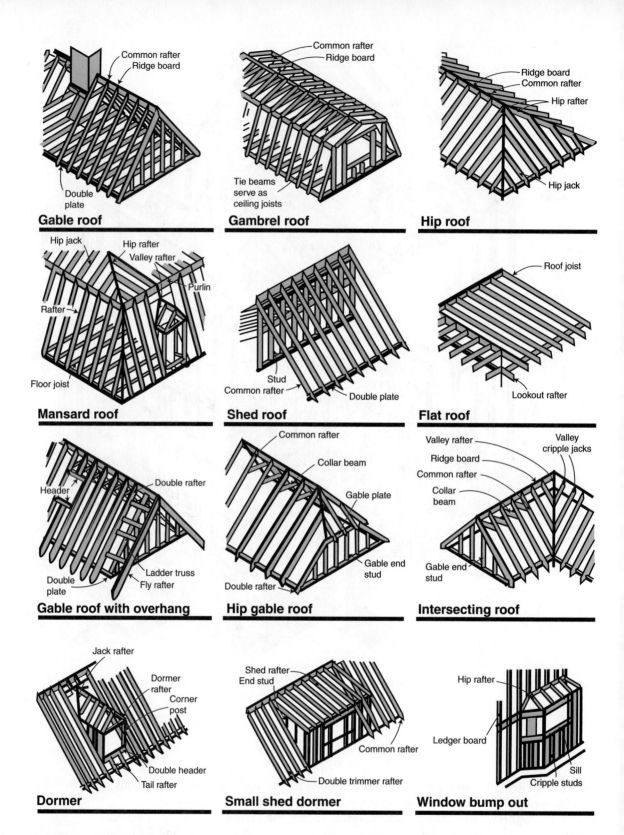

FIGURE 2-5 Typical roof framing plans.

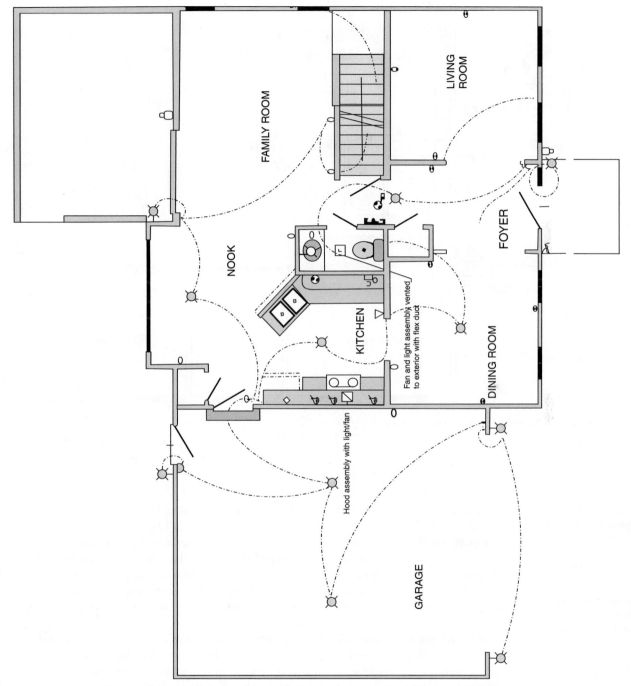

Fan and light assembly vented
to exterior with flex duct

Hood assembly with light/fan

LIVING
ROOM

FAMILY ROOM

NOOK

KITCHEN

FOYER

DINING ROOM

GARAGE

FIGURE 2-6 A utility plan, also called a mechanical plan, for first floor.

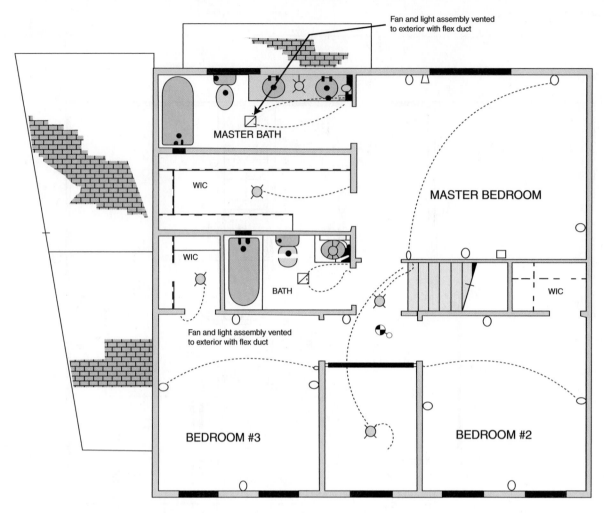

FIGURE 2-7 Second-floor mechanical plan.

Fan and light assembly vented to exterior with flex duct

MASTER BATH

WIC

MASTER BEDROOM

WIC

BATH

WIC

Fan and light assembly vented to exterior with flex duct

BEDROOM #3

BEDROOM #2

elevations ■ Geometric drawings or projections on a vertical plane showing the external upright parts of a building.

window and door schedules ■ Part of the architectural plans. Lists all doors and windows by location, size, and style.

sections ■ Architectural drawings showing a part of a building cut vertically or horizontally so as to show the interior or profile.

these plans (see Figures 2-9 and 2-10). It is equally essential to know and understand the symbols used for the electrical plans (see Figures 2-11 and 2-12).

Elevations show the front, rear, and sides of a structure projected on vertical planes parallel to the planes of the sides (see Figure 2-13). They also give important vertical dimensions. These views show the perpendicular distance from the finished floor to the top of a wall or the underside of a ceiling. The plan also indicates sill heights and countertop and cabinet heights, and can also dimension the centerlines of windows or other openings in a structure.

Window and door schedules list by location what type of assembly is required and its dimensions. **Sections** are views of a cross section cut by a vertical plane (see Figure 2-14). Wall sections are extremely important as they show all the various components, sizes, and configurations and their relationships to each other. For example, the section will show the footing, wall components' sizes, the position of floors and/or joists, the sills for windows, and header heights for windows and doors.

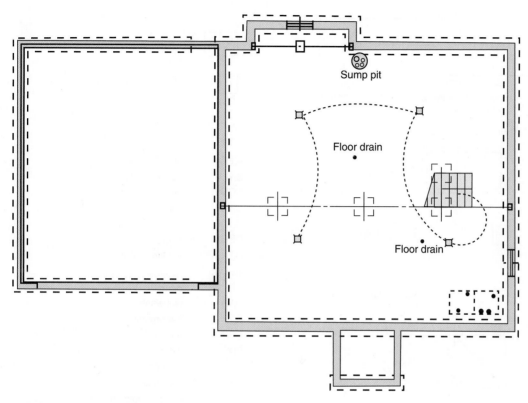

FIGURE 2-8 Foundation mechanical plan.

As-built drawings are extremely valuable to an on-scene commander. They represent exactly how the building and all of its systems are installed; not necessarily as they were intended, but as where they do, in fact, exist. The configuration and sizes of systems are also marked on these drawings. As-built drawings can be created from the drawings amended as the building was constructed. In most cases, however, they are created by a firm specializing in this work, one that meticulously measures and depicts the building and its contents.

Details are drawings done on a larger scale than building drawings and show features that are too small to be seen or do not appear at all at a reduced scale. Wall sections can be details as well. Framing details at windows, doors, or cornices are commonly drawn as sections. This may appear confusing but, like anything else, practice makes you proficient.

It is important to be familiar with the following common terms.

The main parts of a structure are the *load-bearing structural members,* which support and transfer the loads on the structure while remaining in equilibrium with each other. The places where members are connected to other members are called **joints**. The sum total of the load supported by the structural members at a particular instant is equal to the total dead load plus the live load. The total deadload is the total weight of the structure, which gradually increases as the structure rises and remains constant once it is completed.

as-built drawings ■ Drawings showing conditions as they currently are, or "as is."

details ■ Drawings showing the positions and parts of a specific assembly at a specified point.

joints ■ The intersection of two planes or two parts of an assembly.

FIGURE 2-9 Symbols used on plumbing plans. *Source: Dagostino, Frank R.; Feigenbaum, Leslie, Estimating in Building Construction, 6th Edition, © 2003, p. 312. Reprinted by permission of Pearson Education, Inc., Upper Saddle River, NJ.*

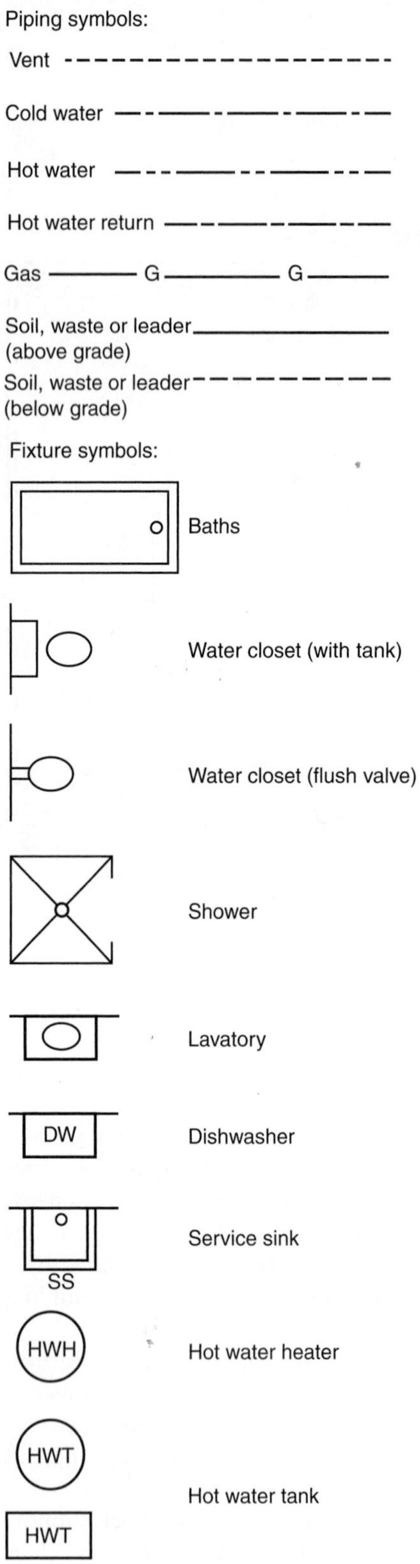

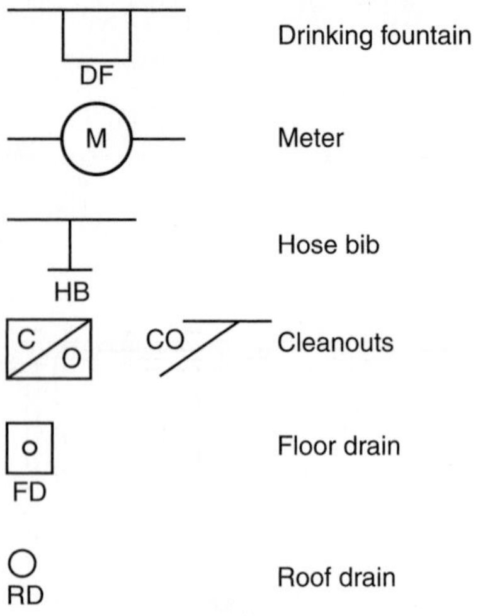

Piping symbols:

Vent — — — — — — — — — —

Cold water — · — · — · — · —

Hot water — — · — — · — — ·

Hot water return — — · — — · —

Gas — G — G —

Soil, waste or leader (above grade) ——————

Soil, waste or leader (below grade) — — — —

Fixture symbols:

Baths

Water closet (with tank)

Water closet (flush valve)

Shower

Lavatory

DW   Dishwasher

SS   Service sink

HWH   Hot water heater

HWT   Hot water tank

HWT

DF   Drinking fountain

M   Meter

HB   Hose bib

C/O   CO   Cleanouts

FD   Floor drain

RD   Roof drain

A.F.D.	area floor drain
B.W.V.	backwater valve
CODP.	deck plate cleanout
C.W.	cold water
C.W.R.	cold water return
DEG.	degree
D.F.	drinking fountain
D.H.W.	domestic hot water
DR.	drain
D.W.	dishwasher
F.	Fahrenheit
FDR.	feeder
FIXT.	fixture
F.D.	floor drain
F.H.	fire hose
F.E.	fire extinguisher unit
H.W.	hot water
H.W.C.	hot water circulating line
H.W.R.	hot water reserve
H.W.S.	hot water supply
H.W.P.	hot water pump
I.D.	inside diameter
LAV.	lavatory
LDR.	leader
O.D.	outside diameter
(R)	roughing only
R.D.	roof drain
S.C.	sill cock
S.S.	service sink
TOIL.	toilet
UR.	urinal
V.	vent
W.C.	water closet
W.H.	wall hydrant

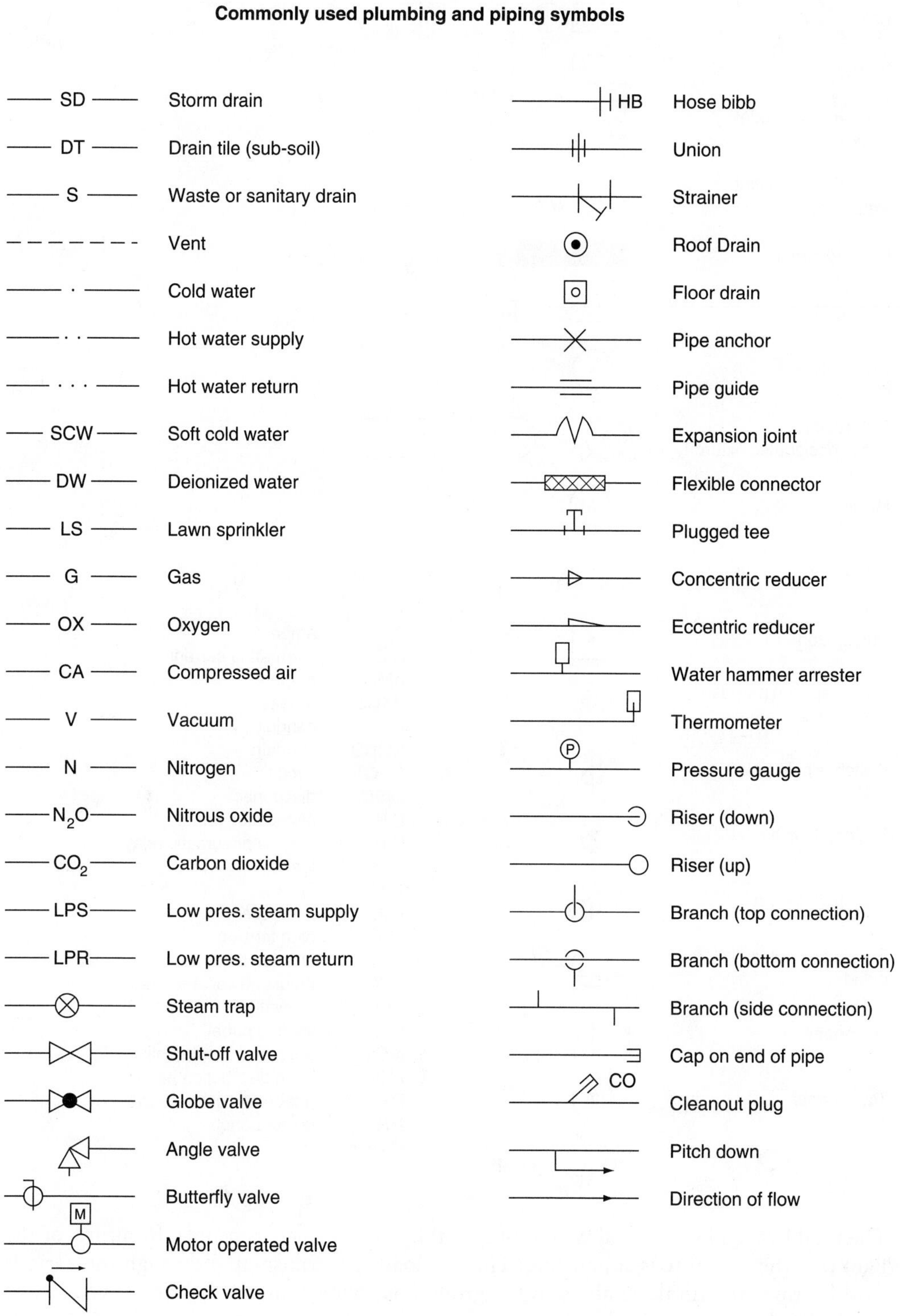

Commonly used plumbing and piping symbols

—— SD ——	Storm drain		——⊣ HB	Hose bibb
—— DT ——	Drain tile (sub-soil)		——╫——	Union
—— S ——	Waste or sanitary drain		——⋉——	Strainer
– – – – – –	Vent		⊙	Roof Drain
—— · ——	Cold water		☐	Floor drain
—— · · ——	Hot water supply		——✕——	Pipe anchor
—— · · · ——	Hot water return		══	Pipe guide
—— SCW ——	Soft cold water		—⋀—	Expansion joint
—— DW ——	Deionized water		—⬚⬚⬚⬚—	Flexible connector
—— LS ——	Lawn sprinkler		—T—	Plugged tee
—— G ——	Gas		—▷—	Concentric reducer
—— OX ——	Oxygen		—◺—	Eccentric reducer
—— CA ——	Compressed air		☐	Water hammer arrester
—— V ——	Vacuum		☐	Thermometer
—— N ——	Nitrogen		℗	Pressure gauge
——N_2O——	Nitrous oxide		—⊃	Riser (down)
——CO_2——	Carbon dioxide		—○	Riser (up)
—— LPS ——	Low pres. steam supply		—⊕—	Branch (top connection)
—— LPR ——	Low pres. steam return		—⊖—	Branch (bottom connection)
—⊗—	Steam trap		—⊥—	Branch (side connection)
—▷◁—	Shut-off valve		——∃	Cap on end of pipe
—▶◀—	Globe valve		—⫽ CO	Cleanout plug
—◁—	Angle valve		—⌐→	Pitch down
—⊕— M	Butterfly valve		——→	Direction of flow
—○—	Motor operated valve			
—▷—	Check valve			

FIGURE 2-10 Graphic symbols used in commercial plans. *Source: Tao, William K. Y.; Janis, Richard R., Mechanical and Electrical Systems in Buildings, 2nd Edition, © 2001, p. 236. Reprinted by permission of Pearson Education, Inc., Upper Saddle River, NJ.*

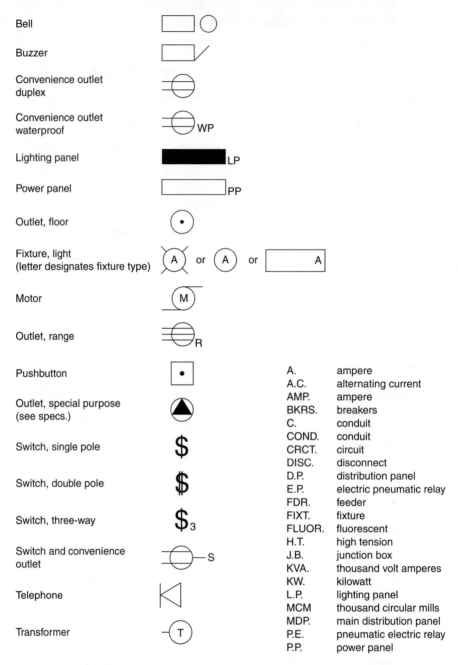

FIGURE 2-11 Residential electrical symbols.

Source: Dagostino, Frank R.; Feigenbaum, Leslie, Estimating in Building Construction, *6th Edition, © 2003, p. 307. Reprinted by permission of Pearson Education, Inc., Upper Saddle River, NJ.*

Bell

Buzzer

Convenience outlet duplex

Convenience outlet waterproof

Lighting panel

Power panel

Outlet, floor

Fixture, light (letter designates fixture type)

Motor

Outlet, range

Pushbutton

Outlet, special purpose (see specs.)

Switch, single pole

Switch, double pole

Switch, three-way

Switch and convenience outlet

Telephone

Transformer

A.	ampere
A.C.	alternating current
AMP.	ampere
BKRS.	breakers
C.	conduit
COND.	conduit
CRCT.	circuit
DISC.	disconnect
D.P.	distribution panel
E.P.	electric pneumatic relay
FDR.	feeder
FIXT.	fixture
FLUOR.	fluorescent
H.T.	high tension
J.B.	junction box
KVA.	thousand volt amperes
KW.	kilowatt
L.P.	lighting panel
MCM	thousand circular mills
MDP.	main distribution panel
P.E.	pneumatic electric relay
P.P.	power panel

The total live load is the total weight of movable objects (such as people, furniture, or the like) that the structure is supporting. The live loads are transmitted through the various load-bearing structural members to the ground as follows: immediate or direct support for the live loads is provided by horizontal members; these are in turn supported by vertical members, which are in turn supported by foundations and/or footings.

Finally, the ability of the ground to support a load is called the *soil-bearing capacity*; it is determined by test and measured in pounds per square foot. Soil-bearing capacity varies considerably with different types of soil, and a soil of a given bearing capacity will bear a heavier load on a wide foundation or footing than on a narrow one. This condition is controlled by a regulating agency but, surprisingly, it is not until ground is broken

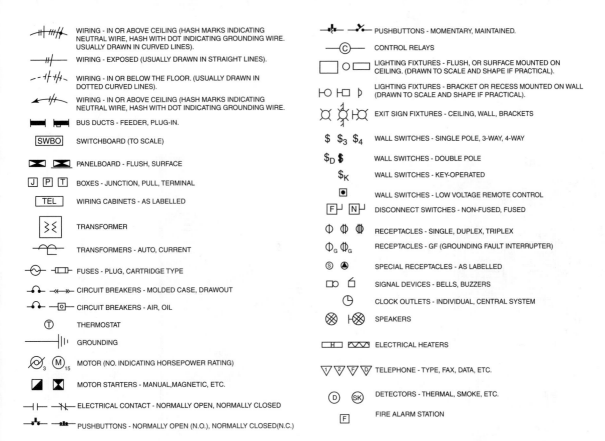

FIGURE 2-12 Electrical symbols used for commercial plans. *Source: Tao, William K. Y.; Janis, Richard R.,* Mechanical and Electrical Systems in Buildings, *3rd Edition, © 2005, p. 423. Reprinted by permission of Pearson Education, Inc., Upper Saddle River, NJ.*

that confirmation is secured. This can be sticky for the builder, engineer, officials, and the owners.

Vertical structural members are high-strength columns; they are sometimes called *pillars* in buildings. Outside wall columns and inside bottom floor columns usually rest directly on footings. Outside wall columns usually extend from the footing or foundation to the roof line. Inside bottom-floor columns extend upward from footings or foundations to horizontal members, which in turn support the first floor. Upper-floor columns are usually located directly over lower-floor columns. A pier in building construction might be called a *short column*. It may rest directly on a footing or it may be simply set or driven into the ground. Building piers usually support the lowermost horizontal structural members.

The chief vertical structural members in light-frame construction are called **studs**. They are supported on horizontal members called **sills** or **sole plates** and are topped by horizontal members called **top plates** or *rafter plates*. Corner posts are enlarged studs located at the building corners or where partitions intersect with perpendicular running walls. In early full-frame construction a corner post was usually a solid piece of larger timber. In today's assemblies the posts are constructed by marrying various pieces of material together to form a wider unit. This is done for increased stability and also to provide an attachment surface for wall, ceiling, or other covering.

Horizontal structural members support weight in a longitudinal axis. A horizontal member that spans a space and is supported at both ends is called a **beam**. A member that

studs ■ In building, an upright member of either wood or steel used in the construction of a wall.

sills ■ The horizontal members that form the lowest members of a frame supporting the superstructure of a house, bridge, or other structure.

sole plates ■ The lowest horizontal member placed on the subflooring upon which the wall and partition studs rest.

top plates ■ In building, the horizontal member nailed to the top of the partition studding.

beam ■ Principal horizontal member used to support a load from post to post or from terminal ends. Used to support floors of a building.

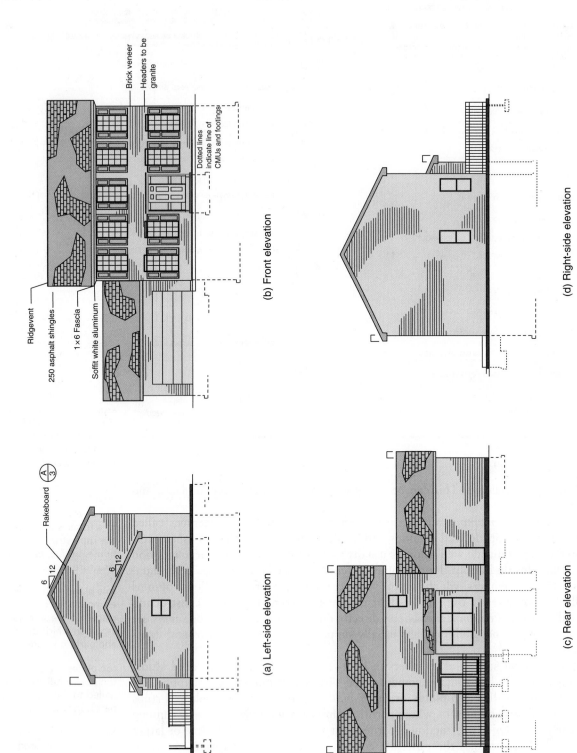

FIGURE 2-13 Elevations: (a) left-side elevation; (b) front elevation; (c) rear elevation; (d) right-side elevation.

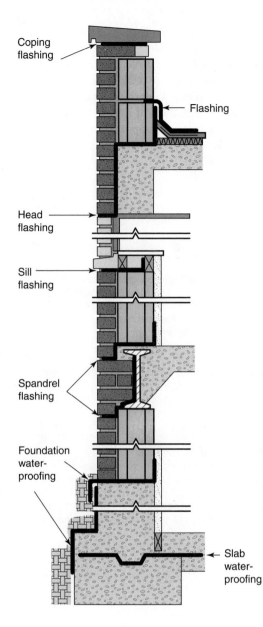

Coping flashing

Flashing

Head flashing

Sill flashing

Spandrel flashing

Foundation water-proofing

Slab water-proofing

FIGURE 2-14 Section view of typical wall showing flashing locations. *Source: Dagostino, Frank R.; Feigenbaum, Leslie,* Estimating in Building Construction, *6th Edition, © 2003, p. 181. Reprinted by permission of Pearson Education, Inc., Upper Saddle River, NJ.*

cantilever ▪ A projecting beam supported only at one end.

open-web joist ▪ Term used in the building industry to describe trusses used as floor members.

bar-steel joist ▪ Term applied to steel truss.

girts ▪ A horizontal member used in braced-frame construction that carries the second floor joists.

girders ▪ A large, supporting horizontal member used to support walls or joists. The term is used for steel or wooden members that support the superstructure of a building.

rafter plates ▪ The framing member upon which the rafters rest. It is usually the top plate of a bearing wall.

lintels ▪ Pieces of wood, steel, or stone placed horizontally across the top of a door or window opening to support the walls immediately above the openings.

is fixed at only one end is called a **cantilever**. Steel members consisting of a solid piece of the regular steel shapes are called *beams*, but a type of steel member that is actually a truss is called an **open-web joist** or **bar-steel joist**. Horizontal structural members that support the ends of floor beams or joists in wood frame construction are called *sills*, **girts**, or **girders**, depending on the type of framing being done and the location of the member in the structure. Horizontal members that support studs are called *sill* or *sole plates*. Horizontal members that support the wall ends of rafters are called **rafter plates**. Horizontal members that assume the weight of concrete or masonry walls over windows and doors are called **lintels**. A beam of given strength, without intermediate vertical supports below, can support a given load only up to a certain maximum span. If the span is wider than this maximum, intermediate supports such as a column or wall must be provided for the wall. When open space is desired without the use of intermediate supports, trusses can be used. Trusses can be wood or metal depending on the type of construction. Instead of mass, the truss utilizes the rules of weight sharing through the engineered configuration;

rafters ■ Sloping roof members that support the roof covering, extending from the ridge to the eaves.

ridge ■ A horizontal timber at the upper end of the common rafters to which the upper ends of the rafters are nailed.

ridge pole ■ The horizontal member or timber at the top of the roof that receives the upper ends of the rafters.

ridge board ■ A horizontal timber at the upper end of common rafters to which the rafters are nailed.

ridge piece ■ A horizontal timber at the upper end of common rafters to which the rafters are nailed.

purlins ■ Horizontal timbers supporting the common rafters in roofs; the timbers spanning from truss to truss.

the form used spreads the imposed load. The assembly is dependent on all members remaining intact otherwise failure can occur with sudden and violent results.

The horizontal or inclined members that provide support to a roof are called **rafters**. The lengthwise (right angle to the rafters) member supporting the peak ends of the rafters is called the **ridge, ridge pole, ridge board**, or the **ridge piece**. Lengthwise members other than ridges are called **purlins**. In wood construction the wall ends of rafters are supported on horizontal members called *rafter plates*, which are in turn supported by the outside wall studs. In concrete or masonry construction, the wall ends of rafters may be anchored directly on the walls or on plates bolted directly to the walls.

SCHOOL OF HARD KNOCKS

In order to become a journeyman carpenter, I spent a great amount of time reading plans and blueprints in my latter years of apprentice school. As a foreman and then a superintendent, I relied exclusively on plans to guide the process of building. For instance, I had to develop a sense of how to run plumbing waste lines installed in the ground lines before a structure was framed. I learned that where we placed those lines had to be exact. Such experience is critical when emergencies arrive.

On September 11, 2001, I arrived on the scene of the Pentagon incident approximately two and a half hours after the attack. I completed a size-up risk assessment upon arrival on the scene; this took over twenty minutes. At the collapse zone I was provided with a cut-section view of the structure. Recognizing the building as Type I fire-resistive concrete construction allayed my fears that it would fail. I relayed this information to the chief from Arlington, who was the *ex officio* incident commander. There was a set of floor plans on a card table that became my command post. When I asked the assembled military officers, at the table, if the plans were for my use they replied yes, and I asked them to draw the impact area on the plans. This made it easier to develop better and more effective strategies and tactics for combating the fire and protecting the firefighters under my command.

Being able to read and utilize floor plans and blueprints is essential to recognizing good or bad fire safety practices during all inspections conducted by operational personnel. An incident commander must be conversant with the language of building processes. A fireground commander who knows how to read drawings will be able to direct building personnel in those tasks necessary to have the building's systems augment the firefighting personnel. A plans officer should be assigned to this task to relieve the commander of these duties if the availability of personnel exists.

ON SCENE

A fire department responds to a high-rise structure fire. There are six rooms burning on an upper floor. There is a report of people trapped on the fire floor, the floor where the fire is located.

1. What drawings would you need as the IC to get a clearer picture of the layout of the fire floor?
2. What part of the drawings would you need in order to study the wall material composition of the partitions on the fire floor?
3. If you wanted to shut off the electricity for the fire floor, which drawings would give you the location of the main shutoffs and floor shutoffs?

FIRST RESPONSE

High-Rise Buildings

High-rise buildings are defined by the NFPA as buildings over 75' in height or beyond the reach of the fire department's longest ladder. They are also any building beyond the reach of the longest pre-connected attack hoseline. It is the opinion of the author that high-rise buildings are those at which firefighters are dependent solely on the building for services—water, electricity, light, and ventilation. Depending on their use, they are may be occupied by many hundreds, if not thousands, of civilians,. They can be apartment buildings, offices, or a combination of the two. Most often they are Type I construction. They can have multiple elevator banks servicing different floors and can also have freight elevators. Most likely these elevators are capable of firefighter service.

The problems the fire service encounters with these structures are, first, their size; many have hallways of 100' or more. Height is a factor; the chances for a safe retreat diminish with each floor above ground. On upper floors, using any of the windows as an emergency exit would mean certain death. Plus, in most high-rise structures it may be very difficult to break the glass and, even if you are successful, the dangers posed to personnel and civilians below can be devastating. Rescues of victims are above the twelfth floor will be extremely difficult, if not impossible, if conditions deteriorate because of smoke and heat. Aerial rescues are usually quite slow. Some considerations for strategy and tactics are listed below:

- The use of the incident command system is absolutely mandatory.
- Thorough knowledge of the building's systems and of reading plans will enable an IC to better direct operations.
- Accountability and SCBA management go hand in hand on these buildings.
- Personnel have to be adequately grouped into crews (of at least two engines and a truck company) and their air usage has to be monitored.
- All attack and vent crews must have adequate backup lines and personnel.
- Interior use of large-caliber nozzles and streams should be part of an IC's tactics.
- Command staff (Operations and Logistics) needs to be at least one floor below the fire.

CHAPTER **REVIEW**

Summary

Understanding the building codes for your area will allow you to familiarize yourself with the construction practices to be expected. If you understand the key terms of the building plans you can interact with tradespeople and officials on their terms. Reading plans and blueprints should be learned in the same manner in which we learn to read road maps.

Obviously the bigger, taller, or more complex the structure, the more pages will be required for the prints. Simply put,

the architectural or building plans will show components and assemblies of the structure,

the mechanicals will show the systems within a building,

site plans will show major utilities around a structure underground, and

details and specifications will show specific assemblies or components.

Review Questions

1. Discuss the role of the NFPA, ASTM, and ANSI in code development.
2. What information would you expect to find in the mechanicals?
3. Discuss the International Residential Code for One- and Two-Family Dwellings under three stories.
4. What is a floor plan?
5. Where would you find the water and electrical cutoffs?
6. What would be more important during size-up, the plot plan or the sectional view of the building?
7. To evaluate ventilation during a fire incident, which part or parts of the plans would you need?

Suggested Reading

NFPA 1. 2006. *Uniform Fire Code.* Quincy, MA: National Fire Protection Association.

NFPA 251. 2006. *Methods of Tests of Fire Endurance of Building Construction and Materials.* Quincy, MA: National Fire Protection Association.

Cowan, D. and Kuenster J. 1996. *To Sleep With The Angels: The Story of a Fire.* Chicago: Ivan R. Dee.

Schorow, S. 2005. *The Cocoanut Grove Fire.* Cowan

Smith, J. 2008. *Strategic and Tactical Considerations on the Fireground.* Upper Saddle River, NJ: Pearson Education.

Stein, L. 1962. *The Triangle Fire.* Ithaca, NY: Cornell University Press.

CHAPTER 3

Building Materials: Stone, Masonry, Steel, Concrete, and Wood

KEY TERMS

ashlar, *p. 49*

corbels, *p. 53*

cornice, *p. 53*

courses, *p. 52*

curing, *p. 59*

dry-stacked, *p. 49*

fieldstone, *p. 49*

fluxes, *p. 54*

heat of hydration, *p. 59*

plasticity, *p. 52*

reduction, *p. 55*

refractories, *p. 55*

rubble, *p. 49*

slump, *p. 60*

smelting, *p. 55*

spalling, *p. 61*

wythes, *p. 52*

OBJECTIVES

After reading this chapter, you should be able to:

- Understand the properties of the main materials used in construction.
- Understand the processes used to manufacture these materials.
- Be able to identify building types by visual size-up.

Resource Central

For additional review and practice tests, visit **www.bradybooks.com** and click on Resource Central to access book-specific resources for this text! To access Resource Central, follow directions on the Student Access Card provided with this text. If there is no card, go to **www.bradybooks.com** and follow the Resource Central link to Buy Access from there.

Introduction

In most fire-related building courses, there seems to be one glaring omission: the course does not address building materials. This chapter was written to illustrate how these materials are made or processed so that you will understand their strengths and weaknesses as they are installed in the structure. There are certainly many other products involved in a structure, but the purpose of this book is to whet your appetite for construction studies; therefore, plastics and other materials are not addressed. The interrelationship between dissimilar materials has to be a key factor when you are performing size-up. This has to occur upon first arrival as well as throughout the incident.

FIREHOUSE DISCUSSION

On Father's Day, June 17, 2001, the members of the FDNY, Fire Department of New York, suffered one of their most emotionally trying fires, an incident, that went from routine to heartbreaking in an instant and left three firefighters dead. The alarm came in at 1420 hours for a reported fire at 12-22 Astoria Boulevard in the Queens Borough of New York City. The building housed a hardware store, Long Island General Supply. The structure, built before 1930, was of ordinary construction and two stories in height. The foundation was constructed of stone and the exterior walls were comprised of brick and mortar. The building actually consisted of two structures interconnected on the first floors and cellars. The hardware store occupied the first floor and cellar of both buildings. The two buildings shared a party wall; a fire door between the two cellars was propped open to allow for free travel.

The cellars of both buildings contained a significant amount of hazardous/flammable materials. These materials are listed below:

Building 1

- Propane (14 ounces)   142 cylinders
- Mapp Gas   27 cylinders
- Oxygen   3 cylinders

Building 2

- Acetone   6–12 gallons
- Denatured Alcohol   24 gallons
- E-Z Alcohol   12 gallons
- Spray Paint   350 cans
- Lacquer Thinner   12–24 gallons
- Methyl Ethyl Ketone   24 gallons
- Mineral Spirits   24 gallons
- Naphtha   24 gallons
- Paint Thinner   24 gallons
- Toluene   4 gallons
- Turpentine   24 gallons
- Xylene   4 gallons

The first unit arrived on the scene at 1422 hours. They stretched or advanced lines and began forcible entry. As other units arrived the second floors of both buildings were searched; no victims were found. Units began making headway in the cellar and it appeared that a knockdown would be forthcoming. An explosion was heard followed by a massive blast: the two victims killed on the Bravo side of the building had been performing ventilation. The third victim was trapped under the stairs in cellar 1 under tons of rubble. The IC struck additional alarms; a total of 144 pieces of apparatus responded. These included forty-six engines, thirty-three ladders, sixteen battalion chiefs, two deputy chiefs, all five heavy rescue companies, and seven squad companies. It took only one hour from the time of the explosion until the body of the third firefighter was discovered, but he had already succumbed to asphyxia.

The fire was started apparently by two teens who had knocked over a can of gasoline. The liquid flowed into the cellar area where the pilot flame of a hot water heater located inside building 2 ignited it; there was also an illegally stored propane cylinder in the cellar.

Assuming that you have hardware stores or building supply centers in your area, or that you have visited either one, discuss how compliance with NFPA 1620 Standard for pre-incident planning can prevent this event from reoccurring.

- Should these facilities be required to post placards or a Material Safety Data Sheet (MSDS)?
- Discuss fighting cellar fires and the pros and cons for doing so.
- The blast occurred approximately twenty-six minutes after the arrival of the first units. Consider how much fire was actually involved if it was not extinguished in twenty-six minutes.
- At what time should an aggressive interior attack be halted?
- What safety precautions could be put into place to avoid this type of outcome?
- Look up the chemicals that were stored. Would any of them react to water in a negative manner?

Stone

Stone is the oldest building material. The term *stone* can be used interchangeably with the term *rock*, although rock is usually used to describe the mineral in its natural state. The Egyptians used stones weighing many tons to erect the pyramids and the Romans used stone for many of their buildings. During the next 1000 or so years, in the Middle Ages, castles, cathedrals, and bridges were constructed entirely of stone. The advantage of stone is that it does not require mortar as its weight, shape, and texture allow it to be laid or set without a bonding agent. The downside of a stone building is that it requires mass; therefore, stone walls are very thick. The foundations may be six to eight foot wide; upper stories may be three to four foot thick. The earliest immigrants to this country erected stone walls by taking stones from the earth, found as they prepared their fields for planting, and setting them in place. This type of wall is called **dry-stacked**. Modern stonemasons can set stone with mortar, but because the grout does not show, this is also known as dry-stacking. The stone used is either **fieldstone** or **dimension stone**. Dimension stone is natural stone that has been cut to specific sizes or shapes. Some examples of dimension stone used for building purposes are granite, limestone, marble, sandstone, and slate. Fieldstone is picked up from the surface of the ground while cut stone is quarried or cut from larger deposits. Stone can also be classified by shape. **Rubble** stone is in its natural state as plucked from the earth; the size and shape will vary. **Ashlar** stones are cut at the quarry and are between one and six inches thick; the lengths and shapes will vary. Usually one or more faces will be cut, making this a good type of stone for veneers. Boulders are round stones ranging from the size of a melon to huge rocks that require the use of cranes or bulldozers to move (see Figure 3-1).

GRANITE

Granite is found in great abundance in the earth's continental crust. There is much debate regarding how granite is formed, but it is generally accepted that various minerals become heated to become magma and are then cooled. The resultant rock then migrates up through the earth's crust by the pressure of geologic forces. Sometimes these pressures are so intense that they have forced the rock vertically and they become mountains. Granite is extremely hard and noted for its durability. It can be pink, red, brown, buff, gray, black, or green in color depending on where it was quarried. Many of our monuments are constructed of granite because of its strength and color variety. An example is the Washington Monument in Washington, D.C.; the Civil War halted its construction, thus the base and bottom one-third was built from Tennessee granite and the balance was constructed after the war of granite from New England (see Figure 3-2). A clear line of demarcation can be readily seen at the spot where the change took place.

dry-stacked ■ A masonry wall laid without the appearance of mortar; masonry stacked without mortar.

fieldstone (or dimension stone) ■ Stone in its original form as collected from the field.

rubble ■ Rough, broken stones or bricks used to fill in courses of walls or for other filling; rough, broken stone direct from the quarry.

ashlar ■ Squared stone used in foundations and for the facing of certain types of masonry walls.

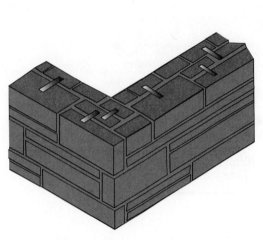

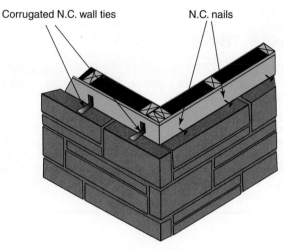

Corrugated N.C. wall ties   N.C. nails

Stone veneer with bond stone anchored to brick backing

Stone veneer anchored to wood frame backing
N.C. = noncorrosive

FIGURE 3-1 Stone veneer walls (ashlar).

FIGURE 3-2 The Washington Monument; a clear line of demarcation can be seen at the spot where the change of granite took place.

Another important aspect of the use of granite is its ability to be reused as part of the green building movement. When a building is demolished (now referred to as deconstructed) the granite can be harvested and reused with very little modification on another project.

LIMESTONE

Another popular stone used in construction is limestone, which is easier to work than granite. Granite is extremely difficult to saw and smooth; limestone is easier and is relatively durable. Limestone is often buff- or cream-colored, but may be reddish or yellowish or have a gray hue. Interestingly enough, limestone needs to season. When first quarried, it often contains groundwater (also called quarry sap). The cream- or buff-colored stone will season in sixty to ninety days. Seasoning produces a lighter, more natural color and allows the groundwater to evaporate. The gray-hued stone takes as long as six months to season because it normally contains more quarry sap. Limestone can be either fieldstone or cut.

Limestone has been found to be susceptible to acid rain. Acid rain occurs when emissions of sulphur dioxide and nitrogen oxides react in the atmosphere with water, oxygen, and other chemicals to form acidic compounds. The resulting chemicals are mild forms of sulphuric and nitric acids.

Limestone is now used primarily as an interior finish for floor tiles, stair treads, and windowsills. Limestone can also be crushed and used in the production of concrete, as a flux for metal refining processes, and as a heat- and weather-resistant coating for asphalt roofing shingles.

SANDSTONE

Sandstone is easier than granite but harder than limestone to work with. Its colors range from pink, buff, and red, to cream, blue-gray, or brown. Sandstone is available as field, quarried, ashlar, or flagstone (cut into thin slabs used for paving).

SLATE

Slate is commonly thought of as a roofing material, but it can also be used for paving; when used as a paver it has a tendency to fracture in layers and split off. An important thing to remember when responding to buildings with slate tile roofing is that employing large-caliber streams to the roof can cause the slate to dislodge and become projectiles that can maim or kill firefighters. Also, slate will become slick when wet or if moss has accumulated.

MARBLE

Marble is also used as a building material. It can adorn walls, floors, or countertops. When polished, its many varieties of hues, veins, and textures make it very suitable for structural and aesthetic uses.

Stone differs from clay or concrete masonry in that it does not need footings. Usually only rock dust is used for the base; rock dust consists of crushed granite that has a granular feel with some lumps. Stone is often used as a veneer, but it needs a lateral structural support as it gets higher than three or four feet. Older structures with stone walls will have thicknesses of two to three feet with mortar joints.

Manufactured stone veneers are replacing natural stone as a siding material. This is especially true in the residential market. This "cultured" stone is composed of lightweight aggregate and air-entrained concrete. The stones are often created using a flexible mold imprinted from natural stone and colored using iron oxide pigments. These faux stones can be applied to any surface without the need for footings. The stone is glued to the substrate using mortar. Almost any type of stone can be replicated. The foundation and chimney of the author's home are covered in faux stone.

Another method for replicating stone is to use the formwork for concrete to imprint the stone or brickwork required. A negative mold is fastened to the forms creating the raised look in the concrete.

Masonry

Humans have been using masonry to construct buildings for centuries. This material outlasts wood under the ravages of nature. Remnants of clay brick, erected sometime around 6000 B.C., can still be found in the Middle East. There are three types of building products associated with masonry: clay brick, concrete block, and stone. Brick and concrete block are the results of a manufacturing process, while stone is installed in its natural state.

CLAY BRICK

plasticity ▪ A complex property involving a combination of qualities of mobility and magnitude of yield that a material can withstand without breaking apart.

Clay brick can be solid or hollow. Clay brick, as its name implies, is comprised mainly of clay. This clay must have enough **plasticity** to be shaped and molded when wet and adequate tensile strength to retain its shape until the units are fired in a kiln. Chemically, the clays used to manufacture brick and its related cousin (terra cotta tile), are compounds of alumina and silica along with differing amounts of metallic oxides and other impurities. The metallic oxides act as fluxes during burning and also influence the color of clay masonry units. The nominal size for clay bricks is 3⅝″ × 7½″ × 2¼″ the weight is approximately 4 pounds per brick.

CONCRETE MASONRY UNITS

Concrete masonry units are made from portland cement, water, aggregates, and, when coloring, water repellency, or other properties are desired, admixtures. This block has also been called cinder block, since blast furnace slag cements, fly ash, and other remnants of manufacturing processes were used before the use of portland cement became cost effective. Lightweight CMU blocks can be manufactured with air-entrained concrete if weight is a factor. Concrete block comes in a variety of sizes, but the most common is 8″ × 8″ × 16″. Concrete cement units can be used alone or as a backup for a clay brick installation. These units can be solid block or hollow core. Solid blocks are used for point load bearing assemblies; for example, to support girders or columns. The hollow core blocks allow for rebar and concrete to fill the voids, creating reinforced masonry. Concrete block units can be used as fire walls, curtain walls, or bearing walls. Clay masonry and concrete block units are stacked along horizontal layers called **courses**, which form vertical layers called **wythes** (see Figure 3-3). Concrete blocks can be faced on one side with a decorative finish when they used alone as a wall material.

courses ▪ Continuous level ranges or rows of brick or masonry throughout the face of a building.

wythes ▪ In masonry, single vertical walls of brick; the single brick-thick part of a wall.

MORTAR

In modern construction the mortar used to connect the units is much more specialized than its nineteenth-century predecessor. Early forms of mortar were predominantly comprised of lime, sand, and water. These have low compressive strengths but do have

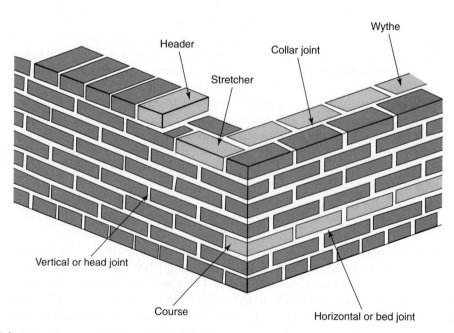

FIGURE 3-3 Elements of a brick wall.

the ability to heal small cracks caused by expansion and contraction. The new forms of mortar types, M, S, N, and O, have different qualities specific to different uses. While O should not be used where freeze/thawing conditions can occur, type M is appropriate for use below grade or where high compressive strengths are needed. Type S is a high-strength mortar, while type N is a medium-strength mortar. Grout is used with hollow core block when solid block strength is needed; this cement-like mix is poured as slurry into the hollow cores or around steel supports to complete the system.

While masonry is indeed strong and enduring, it must be in compression at all times. In an arch all the members are in compression even though they appear to be in tension. Masonry can be incrementally extended outward to form **corbels**, which are often used as a **cornice** or to create elaborate designs; however, it is still always in compression. It is this Catch-22 that can kill or maim responding firefighters who ignore or forget this fact.

MASONRY AND SAFETY

Several basic behaviors should be maintained when firefighters respond to structures containing masonry. First, during any incident where failures have occurred involving floors or the roof, firefighters have to anticipate that a collapse will most likely occur. This is caused by the shifting loads impacting the walls laterally. Masonry walls are excellent in compression but do not stand up well to lateral impacts. For buildings with parapet walls (unreinforced masonry above the roof line, usually resting on a steel lintel or concrete girder), it should be anticipated that the walls will fail if fire is attacking the roof system, or if fires on lower levels are causing load shifts within the structure. For Type III constructed buildings set in a row, the adjoining buildings need to be present as each depends on its partner for support. The mortar used in nineteenth-century masonry applications is especially vulnerable to the effects of hose streams. The water pressure will dissolve the mortar quickly, causing a shift from compression to tension, which is unstable. On modern buildings, the use of single wythe masonry veneers has become common. These assemblies are attached in shear to the framed structure with the use of lightweight metal ties. A significant fire involving the wood framing can cause the veneer wall to fail. The failure does not have to be total for it to be disastrous. A concrete block weighs twenty-five to thirty-five pounds while a nominal-sized clay brick weighs approximately four pounds. If these fall from two or three stories up, they can cause injury or death.

Second, check the duration of the incident. The longer that fire has been impacting the masonry, the hotter the unit and system have become. Masonry acts as a heat sink. It can withstand high temperatures for a period of time depending on whether it is hollow or solid, if grout has been installed in the cores, or if the mortar has been mixed and applied properly. But if the fire effect lasts long enough the masonry will expand, forming hairline cracks that can be significant as the masonry cools; the contraction after expansion can be enough to cause a catastrophic failure. This is readily apparent when a building has a masonry chimney; if the building is lost, the chimney will collapse in just a matter of time.

STRATEGIES FOR SAFETY

So how do we stay safe? First, recognize the dangers of falling masonry. If you are forced to advance a line into a burning structure that contains a parapet wall, consider the risk if the fire is attacking the roof or the underside of the parapet. Your exit could be an ambush. If you are using heavy-duty streams to combat the fire, establish collapse zones. There is still considerable debate among many fire service experts on whether to extend the zone one and a half or two times the height of the building. One thought is to set up the collapse zone and ensure that apparatus marks that line. All personnel should then stay behind the apparatus.

You also need to consider that if you experience a 90-degree collapse, the debris may not stop where it falls; it can continue along the street horizontally at great speeds.

corbels ▪ The stepping out of courses in a masonry wall to form a ledge.

cornice ▪ A term applied to construction under the eaves or where the roof and side walls meet.

Shielding is also important. Firefighters should remain behind apparatus or other solid fixed objects.

When the fire has been knocked down, give the building time to settle down; this is especially true at night when vision is impaired. Once search and rescue is finished, the need for overhaul can wait until daylight. If the incident occurs during the day, use external reconnaissance to establish whether it's safe to go in. The best advice is never to take anything for granted. If the building has been severely attacked by fire for an extended period of time, let it go. Respect collapse zones and use them starting from the arrival of the first units on the scene.

A complete size-up will indicate if the incident will escalate and require large-caliber streams. Position aerial apparatus at the corners first. This will allow for total coverage of the ladder pipes without having to shut down streams and relocate apparatus later. Assign engine companies to supply the ladder companies and assign the firefighters to set up unmanned deck guns or monitor nozzles. Remember, the masonry has been subject to the laws of gravity since it was installed and the fire may have been its undoing. Always put the safety of your firefighters first.

Steel

The earliest known use of iron, the main component in steel, was in 4000 B.C. Its primary use was to make small hand tools. Although 5 percent of the earth's crust is comprised of iron, it is rarely found in metallic form in nature. Pure metallic iron is inherently soft, ductile, and easily shaped, but too weak for most uses. The chemical element, ferrum, provides the term for most iron-based alloys, ferrous metals. Modern iron and steel used in construction get their strength and hardness from the element carbon, however, most steels derive the best combination of properties when carbon is present in amounts less than 1.20 percent. This may appear contradictory, but most steels that are high in carbon content become so hard and brittle that they cannot be readily shaped by current production processes.

Two early types of steel making were *cast* and *wrought*. Cast iron is made simply by pouring raw materials into molds and leaving it to harden. Wrought iron is pulled or drawn. Ferrous metals with low carbon content (less than 0.10 percent) mixed with small amounts of slag are called wrought iron. Early steel-making processes could not sufficiently control the impurities and carbon content to produce metal with the properties of modern steel. Therefore, most early steel was shaped by hand-working or casting. Around 1855 the Bessemer converter process was introduced allowing for mass production of steel. The Bessemer process involved loading raw materials at the top of a cylindrical blast furnace, introducing air across the flame front at the bottom, and capturing the molten material. The process allowed for more uniformity of the finished product.

The primary raw materials needed to produce steel are iron ores, fuels, fluxes, and refractories. Eighty percent of iron ore is mined in the Lake Superior region of the United States. The most common iron minerals occurring in nature, hematite, magnesite, limonite, and siderite, have iron contents up to 70 percent. Hematite is the most widely used in steel making.

The three main fuels used in making steel are coal, oil, and gas. Coal makes up 80 percent of the industry's requirements for heat and energy. The chief fuel of the blast furnace is coke, produced in coking ovens from selected types of bituminous coal. Most coking-quality coal is mined in Kentucky, Pennsylvania, Alabama, and West Virginia. In addition to providing heat and carbon, coke also acts as the reducing agent that separates iron from its oxide in the ores, resulting in the creation of pig iron.

Fluxes are minerals that possess an affinity for the impurities in iron ore or iron. The fluxes grab the impurities and float them off the molten material to form slag. Fluxes are classified as either acids or basics depending on the impurities they are to remove.

fluxes ■ Minerals that, due to their affinity with the impurities in iron ore, are used in iron and steel making to separate impurities in the form of molten slag.

Limestone and dolomite are basic fluxes and sand, gravel, and quartz rock are examples of acid fluxes. **Refractories** are nonmetallic materials used to line steel-making furnaces, flues, and vessels. The most common refractory used is ganister, a form of quartzite rock obtained from quarries and mines in Alabama, Pennsylvania, Utah, California, and Wisconsin.

IRON MANUFACTURING

Iron is manufactured in several ways: in a blast furnace or foundry or by the Aston process. **Smelting** is the metallurgical operation that heats the metal and separates it from impurities. The separation of iron and oxygen is termed **reduction.** The operation of the blast furnace must be continuous in order to maintain the heat. The furnace, usually a cylindrical tower lined with refractories and encased in a steel shell, takes a charge (load) of ore, coke, and limestone, which is loaded through the top. Air is introduced near the bottom. Oxygen in the air combines with the carbon in the heated coke, causing combustion. The molten crude iron (pig iron) is collected and routed to mixers, which are holding furnaces that maintain the molten state of the metal for further routing to steel-making processes. Some metal is cast into forms to make pigs, or *ingots*. (The term *pigs* came from early observances that the ingots looked like suckling pigs as they lay on the floor of the foundry.) The pigs are stored for shipment or for later use. Pig iron is impure, with phosphorous, sulfur, silicon, and manganese levels that are still quite high. The levels of sulfur are significant. Sulfur impurities are what sank the *Titanic*. Sulfur causes steel to remain brittle when subjected to cold temperatures. It was not that the *Titanic* struck the iceberg, but rather that the weak steel cracked and the hand-driven rivets did not remain intact that caused the sinking.

The foundry method uses scrap metal, alloys, and pig iron. These are combined and cast into sand or loam molds. This is also called iron founding. The additional heating causes the pig iron to lose more of its impurities, although the carbon content is still in the 1%–3% range.

The Aston process involves smelting the iron ores in small blast furnaces, or *cupolas*, lowering the sulfur content by adding caustic soda, routing the pig iron to Bessemer converters to burn out impurities and lower the carbon content, and then mixing the iron with iron-silicate slag to form *sponge balls*. The sponge balls are then pressed into massive rectangular shapes called *blooms*. The blooms are reduced by rolling them into smaller slabs and billets (see Figure 3-4).

Steel-making processes involve lowering the carbon content, removing impurities, and adding alloys to obtain the required strength and other properties. The three main steel-making processes are open-hearth, electric, and pneumatic.

STEEL-MAKING PROCESSES

The term *open-hearth method* refers to the furnace. The saucer-shaped hearth is lined with refractories and swept by flames. The furnace itself is made from masonry. Fuels— gas, oil, and tar—are blown into ports at the ends of the furnace over the hearth. Gaseous oxygen is introduced to speed the melting process. The furnace has two levels. The top is used to charge the raw materials; the lower level is used to draw off the steel at the bottom and ladle off the impurities, or slag, from the surface.

Two electric processes are commonly employed: the electric arc furnace and the induction furnace. The electric arc furnace is shaped liked a large kettle and is lined with refractories. Three electrodes project into the furnace. When high voltage is introduced the resulting arc creates heat that melts the charge. The slag is drawn off and the steel is poured from the kettle into paths that route it to other parts of the mill for further processing. This results in long lines of lava flowing continuously throughout a mill. This method is widely used when processing recycled products. The induction furnace uses a

refractories ■ Nonmetallic materials with superior heat and impact resistance used for lining furnaces, flues, and vessels employed in iron and steel making.

smelting ■ Melting of iron-bearing materials in a blast furnace to separate iron from the impurities with which it is chemically combined or mechanically mixed.

reduction ■ Separation of iron from its oxide by smelting ores in a blast furnace.

FIGURE 3-4 Typical blast furnace. *Source: Marotta, Theodore,* Basic Construction Materials, *7th Edition,* © 2005, p. 222. Reprinted by permission of Pearson Education, Inc., Upper Saddle River, NJ.

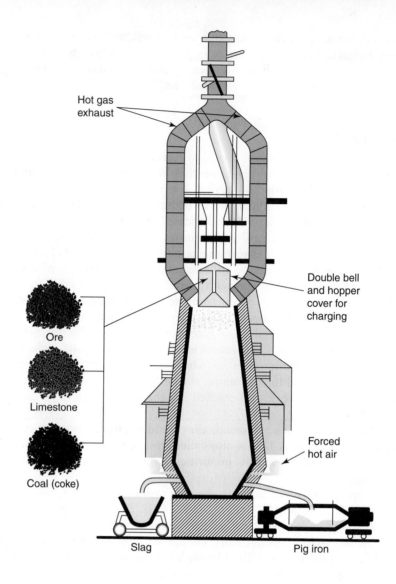

magnesia pot insulated with refractories and surrounded by copper tubing. The windings, which are wraps of copper around an armature to conduct electricity, can be classified in two groups: *armature* windings and field windings. The armature winding is the main current-carrying winding in which the *electromotive force* (emf) or counter-emf of rotation is induced. The current in the armature winding is known as the armature current. The field winding produces the magnetic field in the machine. The current in the field winding is known as the field or exciting current, which is then energized and the resultant heat causes the charge to melt.

There are also two pneumatic processes currently utilized: the Bessemer converter and the basic oxygen furnace. These both operate in the same way except that the Bessemer converter uses air while the oxygen furnace uses oxygen. No fuel is required for these processes because the oxygen is sufficient to create combustion, which in turn removes the impurities and lowers the carbon content. All these processes result in the basic physical properties that differentiate steel from iron: Steel is malleable, it can deform without breaking, and it can be hardened through the use of hot and cold when it is shaped.

Alloys are introduced to molten steel to give or augment its abilities. The alloying elements give the term to the resultant steel product. For example, manganese-molybdenum steels are known for strength and resistance to fatigue.

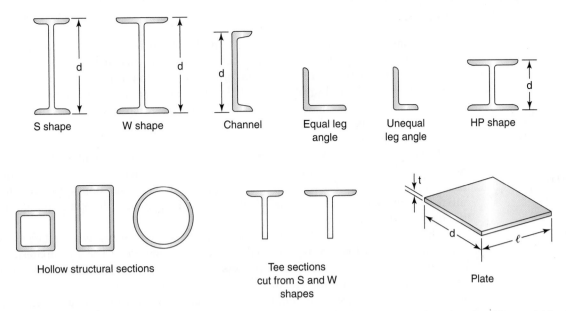

FIGURE 3-5 Various steel shapes used in construction. *Source: Andres, Cameron K.; Smith, Ronald C.,* Principles and Practices of Commercial Construction, *7th Edition, © 2004, p. 336. Reprinted by permission of Pearson Education, Inc., Upper Saddle River, NJ.*

Steel is shaped by several different processes: cast, hot formed, cold rolled, forged, and wrought. Cast steel is ladled into forms and left to harden in place. Much of the automobile industry gets its parts this way. The hot-formed process takes molten steel and draws it back and forth under pressure between a form to create the desired shape. Many items in the construction industry, such as I-beams and H-columns are created this way (see Figure 3-5).

The cold-rolled process takes room-temperature steel and rolls it under pressure to create a desired shape. Forged steel is molten steel sandwiched between two halves of a mold; when the two halves are placed together the resultant pressure produces the finished shape. Wrought steel is used for wire and tube making. This process takes heated steel and continues to pull or draw it into shape. In this process the internal fibers of the steel are continuously stretched to create the desired size, shape, and strength.

BUILDING GREEN

Steel as a roofing material is making a strong comeback and lightweight steel framing members are being used more and more in residential construction. This resurgence in the use of metal has come about due to the recycling process. Many of today's metal building products were yesterday's cars or discarded implements from everyday life.

Concrete Construction

Concrete is used in almost every type and size of architectural and engineering structure. Its numerous advantages make it one of the most economical, versatile, and universally used construction materials available. It is commonly used for buildings, bridges, sewers, culverts, foundations, footings, piers, abutments, retaining walls, and pavements. In residential construction, its primary uses are as foundation walls and footings in the substructure of the building and in flat concrete work, including floor and garage slabs, exterior stairs, and terraces. The plastic quality of concrete—its ability to be readily molded and easily placed into position—offers the opportunity of almost unlimited possibilities in form, pattern, and texture.

HISTORY

Concrete has been utilized for centuries. The Egyptian recipe used gypsum, lime, and water. The Romans perfected a form of concrete to construct their aqueducts and major buildings, many of which remain. They utilized a volcanic rock known as pozzolan in the mix; when water and sand were added, this mixture set up into a hard product. One of the most famous Roman creations is the Pantheon, a large circular structure with a domed roof approximately 143' in diameter. Unfortunately, the Roman recipe for this concrete vanished when their civilization ended.

The English were the next to work with concrete. During the eighteenth century, John Sneeten used clay and limestone that he ground up and then cooked. This mixture was used to join the stone for the Eddystone Lighthouse; what was important was that this mixture hardened under water. In 1824, another Englishman, Joseph Aspdin, patented the name portland cement.

The first concrete manufacturing plant in the United States was in Lehigh, Pennsylvania, and was known as the Coakley Cement Company. The raw materials were cooked in large six-story kilns. This process was labor intensive because the kilns had to be manually loaded and unloaded. In the late nineteenth century the use of rotary kilns made possible the constant manufacturing of cement by allowing the raw materials to be constantly loaded into the kiln; the materials pass through the heating chambers and fall out at the other end as finished clinker.

PORTLAND CEMENT

The essential ingredients of concrete are cement and water, which react chemically in a process called hydration to form another material having useful strength. Natural cement is produced by the grinding and calcination of a natural cement rock, a clay limestone mixture containing up to 25 percent clay. Natural cement is normally yellow to brown in color. The tensile strength and compressive strength of natural cement mortars are low; they have from one third to one half of the strength of normal portland cement. Natural cement was used extensively until the late 1930s, but its quality is variable and it is used less often today. Two types of this cement are available: Type N natural cement and Type NA air-entraining natural cement. Natural cement is used in the preparation of masonry cements for mortar and in combination with portland cement for use in concrete mixtures.

Portland cement is a finely pulverized material consisting principally of compounds of lime, silica, alumina, and iron. Other raw materials can also be used to create cement, including limestone, cement rock, oyster shells, coquina shells, marl, clay, shale, silica sand, and iron ore; when crushed they will also provide the principal compounds. It is said that flour is to cake what portland cement is to concrete. Portland cement is manufactured from selected materials under closely controlled processes. The process starts by mixing proper proportions of limestone or marl with other ingredients such as clay, shale, or blast furnace slag and then burning this mixture in a rotary kiln at a temperature of 2700°F to form a clinker. The clinker is cooled and then pulverized with a small amount of added gypsum. It is ground so fine that nearly all of it will pass through a sieve with 40,000 openings per square inch.

Different types of portland cements are manufactured to meet physical and chemical requirements necessary for specific purposes. The American Society for Testing and Materials (ASTM) defines five types of portland cement in ASTM C150, *Standard Specifications for Portland Cement*.

ASTM Type I   This type is called normal portland cement. It is a general-purpose cement commonly used in sidewalks, bridges, sewers, railways, and masonry units. It is used in applications that are not subject to sulfates and in which the heat generated by hydration will not cause an abnormal rise in temperature.

ASTM Type II This type is a modified cement used when precautions against moderate sulfate attack are important, such as in drainage structures. Type II generates less heat than Type I. It may be used in structures of considerable size and in which cement of moderate **heat of hydration** will tend to minimize temperature rise; examples of such structures include large piers, heavy abutments, heavy retaining walls, and when the concrete is placed in warm water.

ASTM Type III Type III is a high-early strength cement that provides high strengths at earlier time frames. Concrete made with Type III cement has a seven-day strength compared with the twenty-eight day strength of Type I. This cement is used when the quicker pulling of formwork or the effects of cold weather are factors.

ASTM Type IV This low-heat cement is used when the rate and amount of heat generated must be minimized. This is especially true when building dams. When the Hoover Dam was constructed, miles of pipe were placed within the concrete and cold water was pumped through the **curing** concrete to dissipate the enormous amounts of heat generated.

ASTM Type V This is a sulfate-resistant cement used where the soil or water in contact with the concrete contains high concentrations of sulfate.

heat of hydration ■ Heat evolved by chemical reactions with water, such as that evolved during the setting and hardening (hydration) of portland cement.

curing ■ In mortar and concrete work, the drying and hardening process.

MIXING

Concrete is a synthetic construction material made by mixing cement, fine aggregates (usually sand), coarse aggregates (usually gravel or crushed stone), and water in proper proportions. The product is not concrete unless all four of these ingredients are present. A mixture of cement, sand, and water without coarse aggregate is not concrete; it is mortar or grout. (Never refer to a wall or floor as cement. There is no such thing. Cement by itself will not support weight.)

The fine and coarse aggregates are referred to as inert ingredients; the cement and water are the active ingredients. The inert ingredients and cement are mixed together thoroughly. When you see a loaded concrete truck going down the road with the cylinder revolving, the dry materials are being mixed. As soon as water is added, a chemical reaction between the water and cement takes place. This reaction, called *hydration*, causes the concrete to harden. This chemical reaction causes an exothermic reaction resulting in heat of hydration. It is not the water evaporating that causes the hardening (remember: concrete cures under water); the concrete continues to harden only if moisture is available, which is why in summer you might find a worker watering burlap covering a fresh pour of concrete.

The aggregates provide volume for the pour. It would not be economical to pour a floor or sidewalk using only portland cement. The aggregates mean less cement is needed to make the pour. The water used for the mix should be clean enough to drink. But the water can contain small quantities of iron salts (40,000 ppm), sodium chloride and sodium sulfate (20,000 and 10,000 ppm, respectively), silt up to 2000 ppm, and mineral oil not mixed with vegetable or animal oils. The water should not contain alkalines; inorganic salts such as manganese, tin, zinc, copper, or lead; saltwater except in small amounts; industrial wastewater; algae; or sugar (although sugar can be used to retard or accelerate the set time).

The water–cement ratio dictates the strength of the concrete. Too much or too little water severely weakens the batch. For concrete used in commercial or industrial structures for walls or floors, the engineer normally calculates this. In residential construction the mix is often referred to as *bag mixes*. For example, most basements or garage floors will be a four-bag mix, while footings are commonly six-bag mixes. The batch method used for this process takes into account that a bag of cement weighs 94 pounds and is equivalent to 1 cubic foot by loose volume (see Figure 3-6).

FIGURE 3-6 Weight of
1 cubic yard of concrete.

A common batch for one cubic yard of concrete is:	
Cement	6 bags = 564 lb
Water	34 gals = 285 lb
Fine aggregate	1040 lb
Coarse aggregate	1940 lb
Total	**3829 lb**

slump ■ In concrete work, the relative consistency or stiffness of the fresh concrete mix.

For commercial and industrial pours, the **slump** method is used. The mix is charged into a slump cone made of galvanized metal; its base is 8″ in diameter, the top is 4″ in diameter, and it is 12″ high. The cone is filled, or charged, to the top in three layered applications each one-third the height of the cone. The cone is overturned and the height of the resultant is the slump, or sag, of the mix. This should be done at least five times per batch and is accomplished at the delivery truck (or the mixer if stationary). The concrete should only sag, not separate or segregate. If the concrete crumbles it is oversanded (too much sand) and if it segregates it is undersanded. This slump test is conducted to determine the cohesiveness, workability, and placeability of the mix. For commercial and industrial pours, it is critical that all batches are equal in quality and consistency. As a further check, a core test will be conducted after the concrete has set. A diamond-pointed cutter is used to core sample the pour for uniformity of aggregate displacement and cure of the concrete.

POURING OR PLACING

Concrete is *plastic* when placed. This means it is semiliquid, which allows for the concrete to fill voids and to take the shape of the pour. Extra water is often added to the mix to allow for movement; this must be factored in by the engineers so that the strength of the mix is not compromised. Air can also be entrained at this point. Because it does occupy volume, air entrainment allows for lighter-weight batches requiring less mix. Air entrainment is used when concrete will be exposed to weather; it is also widely used when pouring the floors for high-rise buildings, thus lessening the loads imposed.

Concrete is excellent in compression but has no tensile abilities. For this reason it is placed into forms and allowed to set up, creating a solid. Because of the weight of fresh concrete, approximately 3,800 pounds per cubic yard, the forms used must be adequately supported. This is true for wall, floor, or roof pours. The builder can apply a predetermined finish look to the pour with the use of forms. We will examine this more in depth in Chapter 4.

Long ago builders discovered that concrete needs internal support in order to maintain its need to remain in compression. Early builders utilized cast iron girders. Later steel was utilized in girder or column form. Today most pours are reinforced with wire mesh or steel reinforcing rods. Synthetic fibers can be used for reinforcement in non-structural applications only. Nylon, glass, steel, or polypropolene fibers are usually the medium used.

Wire mesh is designated by its gauge and pattern of spacing. For example, you would use 6 × 6 21-gauge mesh for a garage slab. This equates to .021 steel in six-inch squares. Steel rods are available in eleven bar sizes ranging from ⅜″ to 2¼″ in diameter. The number of the rod equates to the number of eighths of an inch of diameter. For example, a #3 bar or rod is ⅜″ while a #11 bar is 2¼″ in diameter. The mesh or rods are placed in the pour at the longitudinal point where it is determined that the pour wants to shift from compression into tension. Normally this will be half the depth of the pour. But on large pours, such as floors or roofs, there may be multiple layers of reinforcement. The size and placement is dictated by an engineer.

If you are involved with an operation where you have to breach a concrete structural member, floor, roof, or wall, you will need to look at the structural drawings, if time and conditions permit, to ascertain the rebar layout. This will prevent potential catastrophic damage to your equipment or further destabilization of the area where you are working.

FINISHING

The plasticity of concrete allows it to settle to an equalized state within the forms. The top surface is then screeded to allow for leveling to occur. *Screeding* is done by hand on small pours and by machine on larger pours. The process consists of dragging a straight edge along the surface of the fresh pour to begin the leveling process. It knocks down high spots of concrete and shows low levels of concrete, which are filled in by workers. This is performed as the pour is made or shortly thereafter.

Floating occurs next. This process pushes the coarse aggregate down into the pour. Normally a wooden or metal bar is used for this purpose. The top one inch or so of the pour is what we often perceive as the finish. A variety of finishes can be accomplished at this time. Most smooth finishes use a steel trowel or troweling machine. Many outdoor applications are given a broom finish to provide traction. This is accomplished by simply dragging a broom lightly across the surface. Workers can also seed, which is done by distributing pebbles, colored glass, mica, or other materials on the surface to create an ornate look. Concrete can also be made to mimic the look of brick, rock, or other materials by applying a form to the surface; this is referred to as stamping. In many cities, the look of cobblestones is appropriate. Creating this by stamping concrete creates the same appearance in a more cost-effective and better-enduring product.

CURING

After the concrete has been finished, there is little to do except to wait for the heat of hydration to cook off the water in the pour, at which point the concrete reaches its maximum strength. Because the water occupied a volume of the pour the builder installs joints or lines in the pour. These joints allow for the expected cracking in the pour from shrinkage. When a pour abuts the intersection of a wall and floor assembly, expansion joints are installed. These joints are made of a compressible material placed between the intersections. Depending on the type of concrete and the amount of cement, the pour may be walked on within hours.

If you respond to a construction site fire or other emergency, use caution if you see supports, or shoring, for concrete walls, floors, or the roof. This indicates that the concrete is still too green (uncured) to be considered self supporting. Care is particularly needed if the pour has occurred on a higher floor. Any failure of the form support will be catastrophic. Bailey's Crossroads was the scene of such an event on March 2, 1973. Fourteen workers were killed and thirty injured when shores were removed on two lower floors at the same time as concrete was being poured on the floor above.

Another way to determine if you are reasonably safe is to see if HVAC or sprinkler mechanics have begun to hang or install their equipment on a floor; if so, the floor above is safe.

Another problem encountered by firefighters battling in a concrete environment is that concrete spalls. **Spalling** is caused by exposure to high heat conditions (above 300°F) and high-pressure hose streams. It is actually a deterioration of the top layer of finished concrete and causes chunks to dislodge from the concrete assembly. These chunks resemble shrapnel.

spalling ■ Spalling is caused by exposure to high heat conditions and high-pressure hose streams. It is actually a deterioration of the top layer of finished concrete that causes chunks to dislodge from the concrete assembly.

Wood

Wood is the one natural product used in the building industry. It has been used in this country since the first settlers cut trees to create space for their homes, gardens, and roads. Trees are classified in two ways: broadleaf (or deciduous) and needled (or conifer). Most deciduous trees are hardwoods, while all conifers are softwoods. But some hardwoods are soft. For example, the hardwoods poplar and gum are softer than many types of softwood. Douglas fir and southern pine, softwoods, are harder than many hardwoods. Most of the structural wood used in building is softwood—either pine or fir. Hardwoods are used for furniture and specialty uses.

All wood is stronger along its longitudinal axis, which is the line of the wood fibers or growth cells of the maturing tree. The basic structure of wood consists of long narrow tubes or cells (fibers or tracheids) that are about the same diameter as a human hair. Their length varies from about ⅟₂₅″ in hardwoods to about ⅛″ in softwoods. Tiny strands of cellulose make up the walls of the cells, which are held together with natural cement called lignin. It is this cellular structure that makes it possible to drive nails and screws into the wood and accounts for its light weight, low heat transmission factors, and sound absorption qualities. When wood is first cut from a tree it contains a great deal of moisture and must be dried to maintain its stability after installation. This is accomplished either by air drying or kiln drying. Air drying requires the wood to be stacked in layers open to the atmosphere while the water evaporates; this process is very lengthy and unreliable. Kiln drying is better suited for production purposes; the wood is stacked in a kiln and baked to a point where the necessary amount of moisture is removed. Wood that still retains its moisture when installed is referred to as *green*. It is sometimes utilized in the construction of pole barns and homes, but using green wood may result in cracking, warping, or splitting of the wood. See Figure 3-7 for a diagram of how wood is cut from logs to make lumber.

Wood is still very widely used in building despite advances by the plastic and steel industries. The following list of advantages gives wood its place in the construction industry:

- Wood is lightweight by volume and can be easily transported.
- Wood is easily shaped in comparison to other materials.
- Wood will lend itself to decorative coatings more easily than other materials.
- Wood resists acids, rusts, saltwater, and other corrosive agents.
- Wood has a high salvage value.
- Wood can be overhauled, making it possible to undo a particular configuration.

All wood is sized. This is referred to as dimension lumber. The sizing pertains to the length of the piece and its unit size. For example: a standard unit of wood used in basic building is the 2 × 4. It comes in many lengths but its unit size will always be the same. So the carpenter will refer to it as a 2 × 4 × 8, implying that an 8″ long piece of 2 × 4

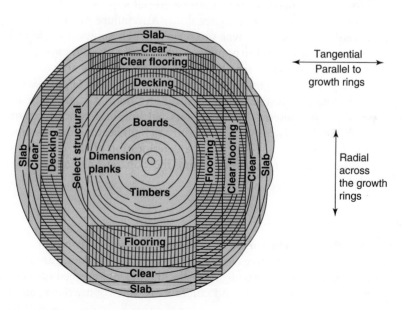

FIGURE 3-7 Lumber cut from logs.

is needed. Wood, when first milled into lumber, is still rough on all four sides. In earlier times this rough lumber was placed into position in its rough-cut stage. The length was cut to fit by the carpenter but the unit size could vary from one piece to the other. Therefore, while most 2 × 4s were in fact a full 2″ × 4″, they could vary by up to ¼″. Today wood is planed to nominal sizes. A 2 × 4 or 2 × 6 or 2 × 8 will be ½″ smaller in width and depth; therefore a 2 × 4 is actually 1½″ × 3½″. At larger dimensions, such as 2 × 10 or 2 × 12, the actual width and depth will be ¾″ smaller.

Wood possesses more compressive strength when erected vertically rather than horizontally. When installed vertically the fibers are running straight up and down. When placed horizontally the fibers will bend, or deflect, and an overloaded piece will collapse in the middle. In addition, all wood members have a natural curvature referred to as the *crown* (see Figure 3-8). For roofs, floors, or stair stringers, the crown should be up. This will allow the unit to straighten out when loaded. For walls and ceilings it is very important that the crown be the same for each member; if not, waves will occur. In many subdivisions, you can look at the end units and see waves in the siding caused by shoddy carpentry. Wood members can share loads if they are placed into position for neighboring, or in alignment to support a load. Most building codes refer to this positioning as spacing. Wood members, whether installed vertically or horizontally, will generally be spaced at 12″, 16″, or 24″ apart, or on center (see Figure 3-9). The term on center applies to the spacing measurement conducted at the center of each member, i.e., if you take a ruler and place it at the center of each vertical or horizontal member the spacing would be on center at generally accepted distances of 12, 16, or 24 inches. This will be different for trusses—each system will be computed by the engineer so you will see 19, 22-1/2, and other distances.

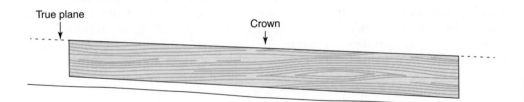

FIGURE 3-8 All dimension lumber has a crown.

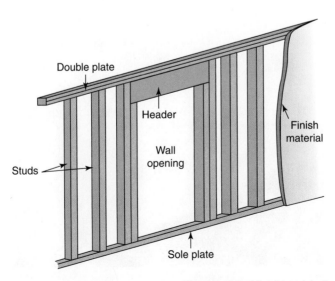

FIGURE 3-9 Wood framing members for a typical wall. *Source: Dagostino, Frank R.; Feigenbaum, Leslie, Estimating in Building Construction, 6th Edition, © 2003, p. 216. Reprinted by permission of Pearson Education, Inc., Upper Saddle River, NJ.*

PLYWOOD

The use of plywood has revolutionized the building industry. Its strength and economy of use allow the builder to achieve stability within the structure's dynamics of combating gravity. When applied horizontally across studs, rafters, or floor joists, the plywood adds strength to the system because of the offset plies within the plywood. When applied vertically, the plywood still adds strength but is vulnerable to sway. It is used for floors, walls, roofs, and as a substructure for cabinets. Plywood consists of glued wood panels made from layers and/or plies of veneer or of veneer and wood. A *ply* is a sheet of veneer cut to thicknesses ranging from as thin as ⅛" to as thick as 5⁄16". The number of layers used in panel construction is always odd: either three, five, or seven. You can count the number of layers on the edges of the plywood sheet. It is produced commercially in 4′ × 8′ or 4′ × 9′ units. It is available in thicknesses from ¼ to 1¼" in ⅛" increments. The sheet may be covered with a *veneer*. Veneers are made from the lower, stronger, portion of a tree. To produce veneers, the log is placed against a sharp blade and peeled like paper from a roll. Often a hardwood veneer such as oak, cherry, or maple will cover a sheet of plywood to give the impression the structure (such as cabinets or a column) is hardwood.

The plywood used for flooring is cut along its edges to form tongue-and-groove joints. The ends are cut straight and flat. Plywood used for floors and roofs is installed perpendicular to the joists or rafters to give maximum strength. For walls, the plywood can be installed either horizontally or vertically, but most often it's installed vertically. Glue is used between the studs, joists, or rafters and the plywood sheet to add strength. Plywood is nailed or screwed into place. Plywood is also used when constructing headers, horizontal members used over openings to redistribute the loads. When used in this manner it should be installed with the grain vertical, but mistakes are often made and it is installed with the grain horizontal, allowing for slight deformation if the span is long.

To avoid waste, a method was developed to reuse the debris left after milling. Flakeboard or particleboard, hardboard, and opposite strand board (OSB) are examples of the materials that can be created from the debris (see Figure 3-10). The leftover chips, flakes, and sawdust are mixed with water and cooked down, and then glue (binder) and paraffin are added. This mixture is pressed together in layers in the same fashion as plywood. These products are used in the same manner as plywood; sheet sizes are also the same. The use of OSB has increased in the past ten years as a cost-savings measure. It is a good product for the building industry, but is highly susceptible to damage from water and/or heat. OSB is dangerous to firefighters because its composition makes it susceptible to failing quickly. It is made from glue, paraffin, and flakes of wood and therefore will fail faster.

FIGURE 3-10 Prefabricated floor joists made with OSB board. *Source: Gary Ombler © Dorling Kindersley.*

SCHOOL OF HARD KNOCKS

I have been part of the processes of the construction and destruction of many buildings. My understanding of the various building materials has enabled me to predict how the materials will behave when affected by fire or after a collapse, as occurred during the incident described below.

My units were fighting a fire inside a Type IV constructed warehouse. The walls were massive stone and mortar. I knew that this type of building would lose its battle against gravity when the main roof members no longer stayed together or the heat built up enough to begin affecting the stone and mortar joints. I set up collapse zones and detailed more safety personnel. When the building came down, it was spectacular, but none of my troops were affected.

On another fire in an auto body repair shop, the first arriving units began to deploy 1½" attack lines. The shop was fairly large, about 100' × 100'. Upon my arrival, I recognized the building was constructed of masonry with lightweight steel bar joists. The units would have to deploy deep into the structure to reach the seat of the fire because the hose streams only afforded a 40 foot reach. No one was placing water on the steel joists. I withdrew the troops and deployed 2½" attack lines, directing the streams for an indirect attack. The fire was extinguished with no danger to the firefighters. Remember: steel begins to move at 1100°F and it will push masonry laterally until the masonry is no longer is in compression. Get water on steel as soon as you can.

FIRST RESPONSE

Commercial Structures

Commercial buildings can be of Type I, II, III, IV, or V construction. Sometimes the businesses in them have been in the same building for generations and the owners may have financial and emotional ties to the community. Firefighters get into trouble with these structures when they underestimate the content fire load, size of the structure, limitations on movement within the building, or the building itself. The only rationale for quick search and rescue is when there are creditable reports of trapped individuals that are accurate as to their location. Even then, backup crews and lines are paramount for a successful outcome. Because many of these buildings have high ceilings (over ten to twelve feet), the heat and smoke conditions may be misleading. Caution must be exercised whenever firefighters are combating incidents in these structures. Limit the personnel in the building until conditions can be verified and then maintain rapid intervention teams (RITs) and accountability. Incident commanders must be ready to pull the plug if conditions warrant.

Some considerations relative to tactics and strategies are listed below:

■ Compliance with NFPA 1620 *Standard for Pre-Incident Planning* should be mandatory when fighting fires in all commercial structures in your area of response.
■ The use of the incident command system is paramount for these responses.
■ Additional safety personnel should be added to monitor all sides of the structure.
■ Size-up reports from the front and rear are needed to make good command decisions.
■ The knowledge of how quickly additional personnel and apparatus can arrive is essential when mounting an attack.
■ If conditions are uncertain, companies should not enter the structure until given orders from command. Little or nothing showing can change quickly!
■ Multiple RIT crews may be necessary if mounting a FAST attack.
■ All units should not lay out to the same water main, certainly not on multiple-alarm incidents.

Fire department units have responded to a fire within a warehouse. The warehouse is six stories in height and measures approximately 150′ × 250′. The foundation appears to be stone and the balance of the exterior brick. There is significant fire, approximately 25 percent involvement, on the second floor.

1. As a company officer, what safety concerns do you have upon arriving on scene?
2. As the safety officer, what predictions must you make regarding failures?
3. As the incident commander, what expectations might you have regarding progress on this incident versus potential collapse injuries? What size lines do you anticipate using?

Summary

Builders use many materials in construction: natural, human-made, and combination products. Stone and wood are examples of natural products. Concrete, steel, and masonry are examples of combination materials. Concrete and masonry are excellent in compression but are terrible in tension; wood has better compressive strength and medium strength in tension. Steel has good compressive and tensile qualities until it is affected by heat. All materials must be installed by craftspeople to ensure quality, and they are all affected by fire and water. Wood will decompose over time. Wood will burn and fail when excessive loads from water and personnel are impinged on it. Stone and masonry will fracture or collapse if mortar is dislodged. Concrete will spall. Spalling is caused by exposure to high heat conditions and high-pressure hose streams; it is actually a deterioration of the top layer of finished concrete that causes chunks to dislodge from the concrete assembly. Steel will eventually lose its tensile strength and melt if exposed to high enough temperatures.

Review Questions

1. Discuss the main differences in strength between steel, concrete, masonry, and wood.
2. Why is stone so unstable under fire conditions?
3. Discuss the differences between cement and grout.
4. What is meant by *crown*?
5. What are admixtures in concrete and what are they used for?
6. What are fluxes used for?
7. Which supplies the strength: concrete or rebar?
8. How many wythes constitute masonry construction?

Suggested Reading

NFPA 1620. 2010. *Standard for Pre-Incident Planning.* Quincy, MA: National Fire Protection Association.

NFPA 5000. 2003.: *Building Construction and Safety Code.* Quincy, MA: National Fire Protection Association.

Feld, J. and Caper, K. 1997. *Construction Failures, 2nd Edition.* New York: John Wiley & Sons.

Graf, D. *Basic Building Data.* New York: Van Nostrand Reinhold Company.

Linde, R. 1987. *The Professional Handbook of Architectural Detailing.* New York: John Wiley & Sons.

www.cdc.gov/niosh/fire-F2001-23. *Hardware store explosion claims the lives of three career firefighters*-New York.

4

Building Components

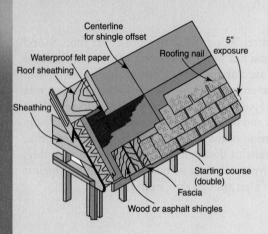

Centerline
for shingle offset

Waterproof felt paper

Roof sheathing

Roofing nail

5" exposure

Sheathing

Starting course
(double)

Fascia

Wood or asphalt shingles

OBJECTIVES

After reading this chapter, you should be able to:

■ Define key building components and their role in the structure.
■ Define key terms associated with a building's construction.
■ Recognize how building components will assist or deter rapid intervention.

Resource**Central**

For additional review and practice tests, visit **www.bradybooks.com** and click on Resource Central to access book-specific resources for this text! To access Resource Central, follow directions on the Student Access Card provided with this text. If there is no card, go to **www.bradybooks.com** and follow the Resource Central link to Buy Access from there.

Introduction

In this chapter we will discuss the various components of the main building's structure. Every firefighter should be able to visualize the building's structural makeup by first looking at the exterior and then confirming once s/he is able to see any interior portion of the structure. This is important for incident commanders as they arrive on the scene and begin to determine whether they have a safe and adequate attack in progress or if they must reengage their resources in another fashion. For example: are you looking at a wood-framed Type V with a one-wythe veneer of brick or stone or are you looking at a masonry building with cement block and clay brick face or a concrete, cast-in-place or tilt-up wall? We will answer these questions in this chapter.

Foundations

Foundations serve as the base for all weight distribution from the roof down to the ground. They must be strong enough to support all the weight above plus be substantial enough to resist lateral forces such as snow loads, wind, and the pressures exerted by the ground itself. The foundation walls are prevented from sinking into the ground by footings. If you watch a construction, the builder digs trenches forming the outline of

foundations ■ That part of a building below the surface of the ground on which the super-structure rests.

FIREHOUSE DISCUSSION

The Southwest Supermarket fire occurred on March 14, 2001 in Phoenix, Arizona. The origin of the fire was a dumpster on the exterior at the rear of the structure. The first units were dispatched at 1654. The fire would kill one firefighter and seriously injure four more before the event was finished.

The building was a one-story L-shaped structure built in 1956 of approximately 27,905 square feet. The foundation was a concrete slab. The building's exterior walls were constructed of unreinforced concrete masonry units (CMUs). The ceiling height to the underside of the metal trusses was twenty feet. The roof consisted of lightweight metal trusses spaced six feet apart with wood, foam (insulation), and a built-up tar and paper assembly. The building had been annexed into the city but was not required to have a sprinkler system.

The first arriving unit reported a fire involving cardboard on the outside of the building, but with exposure to the building. The fire was knocked down quickly and the incident commander ordered crews to check inside the structure for any potential victims and for any extension of the fire. It became apparent to crews on the scene that the fire had extended into the building. The engine company to which the victim was assigned advanced a handline into the structure. The smoke had banked down and visibility was severely limited. There were high heat conditions at the floor level.

The unit had been operating for a short period of time when the victim related to his officer that he was low on air. The officer instructed all members of the unit to "follow the line out of the building." The department's term for this action was, at the time, being *on line* rather than being attached to a rope. The officer and one other firefighter made it to the exterior. The victim and another firefighter became disoriented and ran out of air. The victim radioed his command that he needed assistance.

At this same moment the incident commander had been positioning companies to vacate the building and go defensive, with all units operating outside the structure with larger lines. For approximately the next hour various units had contact with the victim. He was unable to coherently respond to commands and became combative several times. When he was finally retrieved from the building he was unconscious and had no pulse.

This incident led to many changes in the Phoenix Fire Department's response to commercial buildings.

the building's shape and the location of load-bearing walls (see Figure 4-1). The depth of the footings is dictated by local building codes; they must be deep enough to allow for sufficient concrete with rebar to support the load of the structure as well as deep enough to be below the frost line in order to prevent the heaving of the footings. **Heaving**, as it relates to the building process, is concerned with the depth in earth at which the frost line occurs. If the building's footings are above the frost line it is susceptible to the upward pushing force that occurs when ice forms at the frost line and pushes upwards towards the source of moisture, but is restrained by the soil. In most cases, this occurs during the spring thawing period.

Many buildings on the east coast of the United States have basements, so the foundation is excavated deep enough to allow for this. The width of the footings is usually twice the thickness of the foundation wall. The footings must be square and level. In some areas of the country basements are impractical. The ground might be impervious, such as in the southwest; or groundwater too high, such as in the southeast; or the area could be prone to other natural forces, such as earthquakes. In these cases, the builder often combines foundation and footings by using a spread or mat pour. This contiguous pour is often stiff for the footings, creating less slump for what becomes the ground floor.

On the west coast of the United States, or in areas prone to earthquakes, the foundation needs to be securely fastened to the concrete footings. During an earthquake, the movement between the ground and the structure must be kept synergistic. This usually requires rebar extending from the footings through the foundation walls. The structure must also be securely fastened to the foundation walls. The through-bolt method is widely used with bolts spaced approximately one foot apart (see Figure 4-2). When the structure has a knee wall, as do split-level homes, a combination of brackets, plywood framing, and through bolts are often utilized (see Figure 4-3).

Before the pour, plumbers and electricians must install their rough connections and piping in the ground; these must be accurate and plumb. *Plumb* refers to a condition in which the structure, plumbing line, or wall stud is perfectly vertical with no deviation. It also refers to an object being perpendicular to another object that is level or perfectly horizontal.

heaving ■ As it relates to the building process, is concerned with the depth in earth at which the frost line occurs.

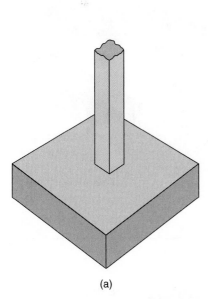

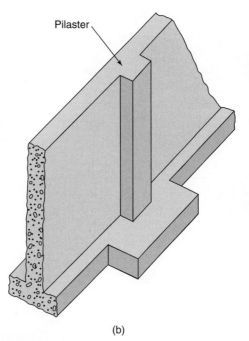

Pilaster

(a)

(b)

FIGURE 4-1 Footings prevent foundation walls from sinking into the ground: (a) isolated spread footing supporting column; (b) continuous spread footing under wall, enlarged at pilaster; (c) typical continuous wall footing.
Source: www.shutterstock.com.
Copyright: prism68.

(c)

   The use of concrete masonry units or poured-in-placed concrete is quite common for residential structures. For commercial structures, especially high-rise buildings, poured concrete with rebar embedded is necessary to carry the loads above.

   Water is almost always present in the ground and must be prevented from entering any subterranean areas of the structure. In residential construction asphalt coatings are used; the asphalt is applied to the block or concrete below grade. If the water table is too high or the soil is compacted the water will not disburse; a sump is often employed to remedy this condition. However, if the water is not diverted away from the foundation the weight of the accumulated water results in lateral pressure that can collapse the wall.

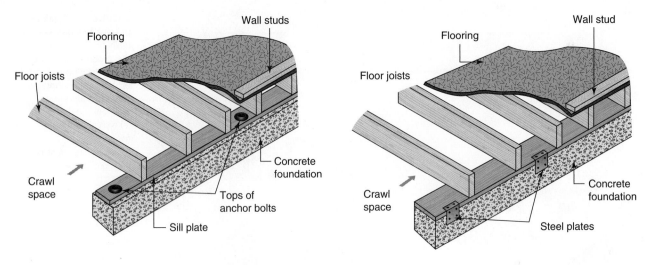

FIGURE 4-2 Through-bolt method for seismic protection.

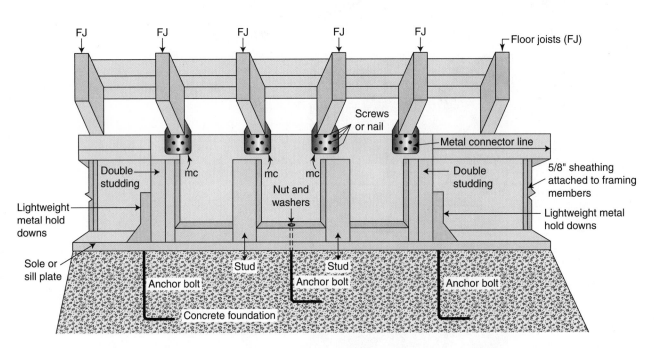

FIGURE 4-3 Residential seismic connection at foundation wall (with cripple well); 5/8" structural sheathing attached to framing.

In commercial structures, the application of grout to the areas between the walls and the surrounding soil is required.

The weight of a structure should be sufficient to hold it tight to the foundation, but in fact it is not. Therefore an attachment connection in the form of bolts and washers, called *anchor bolts*, is used to hold the upper structure to the foundation. To install anchor bolts, the sill plate is drilled and the bolts penetrate through and are attached with the use of a nut and washer. Strangle straps may also be used; strangle straps are made from lightweight steel and are nailed into the sill plate. These connections occur in concrete masonry unit and poured concrete configurations (see Figures 4-4, 4-5, and 4-6).

FIGURE 4-4 Metal straps for anchoring sill plate.
Source: www.shutterstock.com. Copyright: Brandon Bourdages.

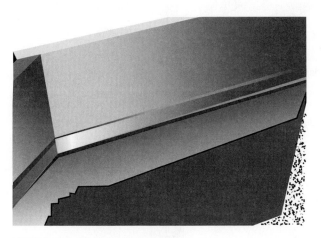

FIGURE 4-5 Sill gasket. *Courtesy of Michael Smith.*

FIGURE 4-6 Ladder-type sill plate. *Courtesy of Michael Smith.*

Walls

The vertical supports for the structure are the **walls**. In Type V balloon-framed structures, the walls run continuously from the foundation to the underside of the roof. Walls form the vertical skeleton of the structure. There are many different terms for walls, including bearing, nonbearing, partition, party, curtain, plumbing, fire, exterior, and interior, and we will look at them individually and collectively.

Wood-constructed walls and lightweight steel stud walls use the same terminology for components. All vertical members are called *studs*. These studs are spaced approximately 12″, 16″, or 24″ apart depending on whether a wall is a bearing wall or not. The rule of thumb for bearing walls is: in residential construction, look for 12″ or 16″ on center; in commercial construction, look at your local codes. The bottom horizontal member is called the sill plate, sole, sole plate or mud sill. The upper horizontal member is called the *top plate*. In most codes. the top plate is required to be doubled for load-bearing walls in order to support and redistribute offset point loads. To keep the lines of assembly even, the top plates of nonbearing walls are also doubled if nominal stock is used. If studs are cut to fit the space, then the top plate can be singled for nonbearing walls. For openings over windows and doors, the weight is transferred horizontally

walls ■ Upright structures of definite dimensions for enclosing space in a building or room.

by headers to king jacks (shortened studs that support the underside of a header), which carry the load vertically. The horizontal member under a window is called a *sill*. It is supported by vertical members called *cripples*. Cripples are also present above a framed door opening. They are placed above a header on the underside of the top plate (see Figure 4-7).

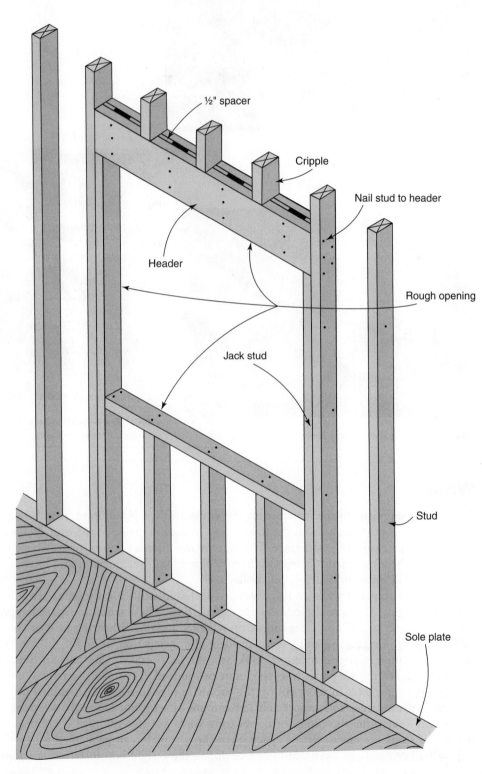

FIGURE 4-7 Wall framing including headers for windows and doors.

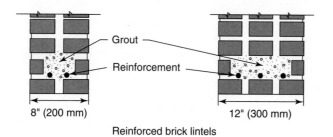

FIGURE 4-8 Brick lintels. *Source: Andres, Cameron K.; Smith, Ronald C.,* Principles and Practices of Commercial Construction, *7th Edition,* © 2004, p. 449. Reprinted by permission of Pearson Education, Inc., Upper Saddle River, NJ.

8" (200 mm) 12" (300 mm)

Grout

Reinforcement

Reinforced brick lintels

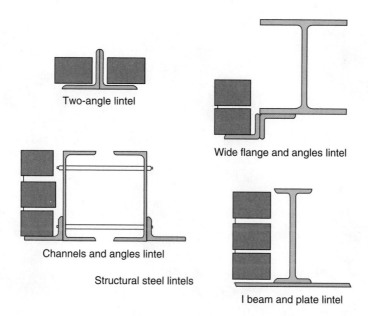

Two-angle lintel

Wide flange and angles lintel

Channels and angles lintel

Structural steel lintels

I beam and plate lintel

Walls constructed of masonry, concrete, or stone share the same name: *lintels*. Lintels, which are normally constructed from steel, carry weight above door and window openings in the structure (see Figure 4-8).

Buildings constructed of solid masonry walls can be identified by looking at the throat space at window and door openings. If the space appears to be at least 8" from the door or window assembly to the outer plane of the masonry, then the building will almost always be solid masonry (see Figure 4-9). However, if the space is approximately 3" or 4", then you are most likely looking at a veneered wall with a single wythe of brick or stone (see Figures 4-10 and 4-11).

TYPES OF WALLS

Bearing walls transfer the weight from the roof to the base of the foundation. These walls must be constructed at a size substantial enough to accomplish this purpose. The sizes and the spacing of the members will be dictated by the building code and certified by a structural engineer. *Nonbearing walls* carry only their own weight. Their purpose is to separate or enclose spaces. The following terms can be applied to nonbearing walls: partition, curtain, or interior. They all mean the same thing. A plumbing wall is usually two partitions set parallel to each other and far apart enough to install the stack and vents for the plumbing system. The builder can also use a wide enough system to dispense with the double wall, but this requires more drilling and cutting by the plumber when running the pipes. An exterior wall is any wall that

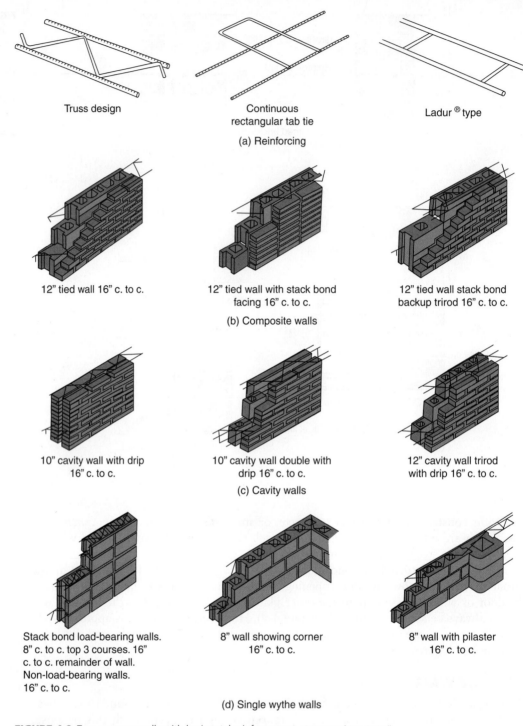

Truss design

Continuous rectangular tab tie

Ladur ® type

(a) Reinforcing

12" tied wall 16" c. to c.

12" tied wall with stack bond facing 16" c. to c.

12" tied wall stack bond backup trirod 16" c. to c.

(b) Composite walls

10" cavity wall with drip 16" c. to c.

10" cavity wall double with drip 16" c. to c.

12" cavity wall trirod with drip 16" c. to c.

(c) Cavity walls

Stack bond load-bearing walls. 8" c. to c. top 3 courses. 16" c. to c. remainder of wall. Non-load-bearing walls. 16" c. to c.

8" wall showing corner 16" c. to c.

8" wall with pilaster 16" c. to c.

(d) Single wythe walls

FIGURE 4-9 True masonry walls with horizontal reinforcement. *Courtesy of Dur-O-Wall.*

has one side exposed to the elements. An interior wall is contained wholly within the building. A firewall is constructed of a noncombustible material and its sole purpose is to prevent the migration of fire and smoke from one space to another, however, it can also serve as an interior nonbearing partition. The terms for wall systems will become familiar with experience. A field study of the wall erection process at a job site can help provide that.

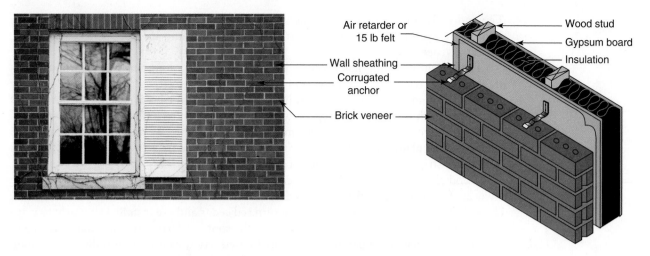

FIGURE 4-10 Brick veneer wall with wood backup. Notice depth of throat. *Source: www.shutterstock.com. Copyright: Maxy M.*

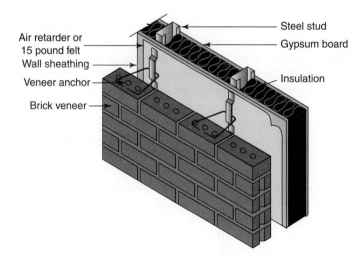

FIGURE 4-11 Brick veneer wall on steel stud backup.

Exterior Wall Coverings

Wall coverings are referred to as exterior or interior. Exterior wall coverings are described as siding or veneer. The purpose of exterior wall coverings is to protect the interior from the effects of weather and to enclose the structure in a watertight membrane. Residential construction will utilize wood, aluminum, vinyl, masonry, or composite materials. Commercial construction will use masonry, metal, or concrete or a combination of masonry, metal, or concrete. For many Type II structures, the front will be made to look attractive, while the sides and rear will most often be metal (either steel or aluminum) installed vertically.

The use of lightweight materials allows for the installation of veneers that appear to be stone, tile, or brick. These can all be applied to the exterior of all construction Types I through V. If you are not knowledgeable regarding these processes, it can be confusing when you are attempting to perform a size-up. And it can be disastrous if you are confronting collapse potentials. As you approach the building, observe the throats of doors and windows. The space between the outside line of the wall and the frame for the window is your clue in identifying a veneered wall versus a true masonry wall. If the throat is more than 8″, it should be true masonry. For any measurement less than 8″, suspect a veneer.

The advent of siding has brought building exteriors a long way. The earliest forms of siding for residential construction involved wood. Early settlers in this country installed the planks of wood horizontally. They started at the bottom of the structure and overlapped each succeeding course (row). This installation technique is called *clapboard*. Most clapboard siding was made from eastern white pine. Wood can also be installed vertically. The planks were installed with a butt joint, side by side, and the joint covered by another strip of wood; this style is called *board and batten*.

Wood shingles and shakes have been used since colonial times. Shingles are usually sawn at a mill, while shakes are hand split from the log. They will vary in size from 12″–48″. They are usually made from western red cedar and are installed similarly to clapboard siding. If they are installed on the roof, the roof is referred to as a cedar shake roof.

The wood was often painted with a lead-based paint, so use caution during overhaul of residences over 100 years old. Be sure to use your SCBA!

Wood is still used today, but cedar, cypress, and redwood have largely replaced white pine. On less expensive residential construction, you may find plywood with vertical grooves cut into the surface to simulate vertical wood siding. This type of siding is called T1-11 or T111.

The use of stucco developed as a result of the need for fireproofing and maintenance-free exteriors. Stucco is installed in two parts. The first layer, or coat, is comprised of cement, water, sand, and lime. The second coat includes gypsum and a mixture of crushed stone and pigments called *stone stucco* or plaster and crushed marble called *scagliola*. The second mixture became prevalent throughout the southern part of the United States; its fireproofing ability was a result of the inert products contained in the mixtures.

The use of metal for siding began in the early twentieth century. By the 1920s, Sears, Roebuck and Co. began offering embossed steel panels simulating brick and stone. There were significant problems with this siding; it was prone to warping, resulting in the structure being vulnerable to rain intrusion. In 1939, Frank Hoess, an Indiana machinist, received a patent for a metal panel with a U-shaped flange that allowed for interlocking the panels, which alleviated the leak issue. Hoess' method is still used today.

In the 1950s aluminum siding made it possible for people to achieve individuality in an economically feasible manner. It was available in a wide array of colors baked on at the factory. It also allowed for the discontinuance of the use of asbestos or asphalt shingles, which had been in vogue up to that point. You may still find asbestos or asphalt shingles in use; many urban areas still have these types of sidings on outbuildings such as sheds or garages. It is best to use SCBAs even while fighting these fires from the outside.

In today's building industry, the use of vinyl siding for residential construction is more economical and gives a better finish because the color is part of the whole product and not baked on. This siding can be installed vertically but most often it is installed horizontally in 12″ wide strips. It is extremely easy to pull off with any flat tool. It adds nothing to the structural stability of a building and thus cannot be factored into size-up.

Siding is attached to the sheathing of the exterior walls. Wood, aluminum, and vinyl siding are attached using either screws or nails. Metal siding is attached using screws or rivets. Sheathing has evolved also. In buildings constructed prior to the advent of plywood, the sheathing will most likely be planking, either rough cut or dimensioned, installed either horizontally or diagonally to the studs. By installing the planking in this fashion, the builder racks in the building, thus preventing vertical shifting of the structure. Most builders today will use plywood or OSB (oriented strand board) on the corners of the structure. This is used to prevent the structure from losing lateral stability. In some instances the entire exterior substrate is plywood or OSB. To prevent racking, 1/2″ aluminum foil-clad foam board or 1/8″ foil-clad cardboard is used with a diagonally installed wire frame.

Some codes will allow combinations of foam, fill-faced cardboard and plywood, or OSB. Check your local codes for what you may expect.

BUILDING GREEN

The purpose of building green is to create structures that are energy efficient and environmentally friendly. The concept of green building is not new. It was first introduced during the 1970s when a fossil fuel shortage occurred in the United States.

The driving political forces behind this movement do not involve the fire service or address the potential for adverse conditions for firefighters battling fires within green structures. We will consider two green elements that can have catastrophic effects for firefighters: SIPs and advanced framing or optimal value engineering (OVE) framing of wooden structures.

STRUCTURAL INSULATED PANELS

Structural Insulated Panels (SIPs) have been in use since the 1930s. Their use was thought to help reduce the harvesting of forest products. SIPs are constructed of foam (expanded polystyrene, extruded polystyrene, or polyurethane) sandwiched between layers of plywood, OSB (which is most commonly used today), or sheet metal. See Figure 4-12 for a typical assembly of a SIP. For below-grade installations, cement board is used. SIPs can be used in wall, floor, or roof assemblies. The connection between foundation and vertical walls is tenuous. A solid sawn member is fastened to the sill; the panels align over the member and are secured with nails or screws. This connection will fail rapidly as movement of the panel (90-degree or inward/outward wall collapse) will pull the screws or nails through the panel. This differs from conventional framing in which each framing member is attached to the panel. See Figure 4-13 for SIPs wall and foundation detail. The depth of SIPs can range from 4½″ for typical walls to 12½″ for typical roofing applications (see Figure 4-14). The size of panels range in sizes from 4″ to 24″ wide by 8′ to 9′ high. The larger panels (usually used for roof assemblies) will require a crane because of their significant weight.

Window and door openings are created by installing sawn members, to which the doors and windows will be attached, into the panels at the openings (see Figure 4-15). During size-up, this process further disguises the use of SIPs.

Structural Insulated Panels (SIPs) ■ SIPs are constructed of foam (expanded polystyrene, extruded polystyrene, or polyurethane) sandwiched between layers of plywood, OSB (which is most commonly used today), or sheet metal.

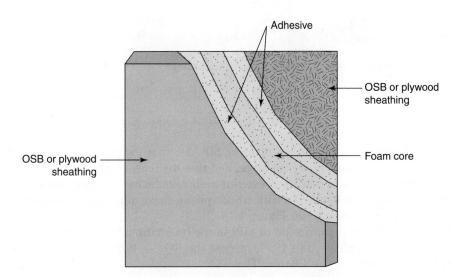

FIGURE 4-12 Typical SIP.

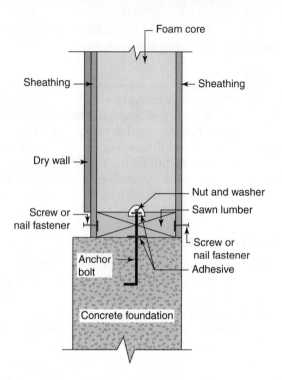

FIGURE 4-13 SIP wall and foundation detail.

Foam core

Sheathing — → Sheathing

Dry wall —

Nut and washer

Sawn lumber

Screw or nail fastener

Screw or nail fastener

Adhesive

Anchor bolt

Concrete foundation

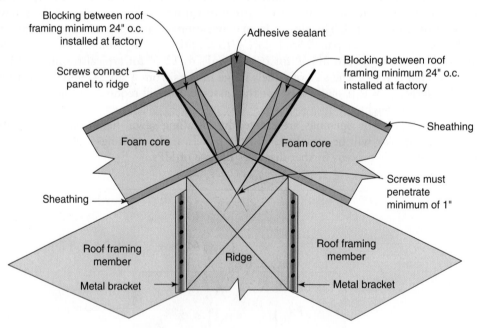

FIGURE 4-14 SIP roof peak detail.

Blocking between roof framing minimum 24" o.c. installed at factory

Adhesive sealant

Blocking between roof framing minimum 24" o.c. installed at factory

Screws connect panel to ridge

Sheathing

Foam core

Foam core

Screws must penetrate minimum of 1"

Sheathing

Roof framing member

Ridge

Roof framing member

Metal bracket

Metal bracket

For the builder, the advantages of using SIPs are that the panels are made off site and trucked to the job site, saving on labor, and that the finished panels serve as framing members, insulation membrane, and exterior siding, reducing the completion time for projects. The panels can be erected with wood splines, using dimensioned lumber or thin wedges connecting the panels (see Figure 4-16).

The ICC approved the inclusion of SIPs in the International Residential Code (IRC) in 2007. The ICC, established in 1994, merged the BOCA, ICBO, and SBO codes into those of the IRC and IBC (see Chapter 1).

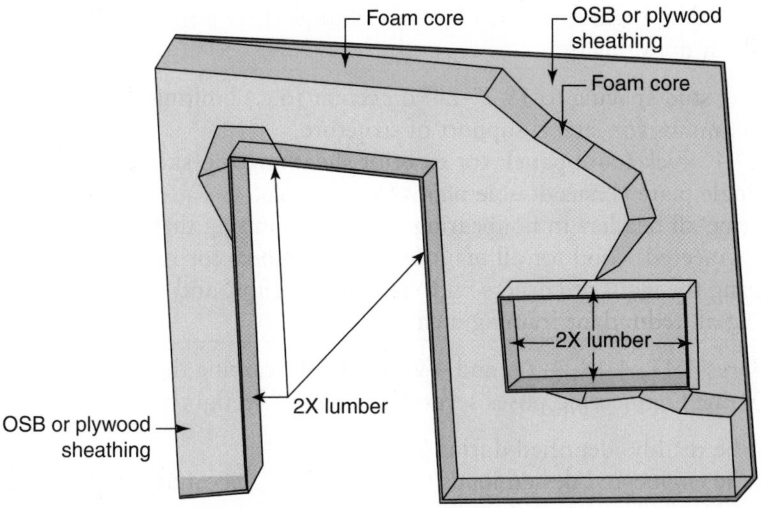

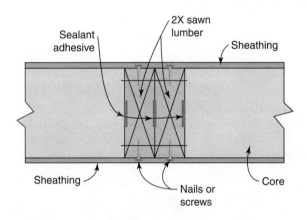

FIGURE 4-15 Panels with openings for windows and doors.

Foam core — OSB or plywood sheathing — Foam core — 2X lumber — OSB or plywood sheathing — 2X lumber

FIGURE 4-16 SIP connections between vertical panels.

Sealant adhesive — 2X sawn lumber — Sheathing — Sheathing — Nails or screws — Core

SIPs can pose several hazards to firefighters at an incident.

- SIPs will not be differentiated during size-up.
- SIPs will lose their structural capabilities quickly when directly impinged by fire as the flames deteriorate the foam core and plywood skin.
- When SIPs are used there is no redundant load bearing framing, only the panels.
- If not installed properly, the foam will degrade over time because of moisture infiltration, making the roof untenable.
- For most residential codes the only fire resistive requirement will be fifteen minutes achieved by one layer of ½″ Sheetrock.

OPTIMAL VALUE ENGINEERING

Optimal Value Engineering (OVE) of a wooden structure is a technique for framing wooden houses, designed to lessen the requirement for wooden building materials, gained momentum in the 1970s. It did not hold much appeal for the industry until the recent green movement. The technique has since garnered attention because:

it reduces the amount of lumber needed to construct a house,
the increased stud spacing allows for more insulation,
the use of engineered lumber reduces waste, and
it lowers the labor cost.

Optimal Value Engineering, OVE, is a technique that removes all superfluous framing materials. It does this by

- increasing stud spacing to 19.2"–24" on center (o.c.) limiting the use of sheathing to code minimums for lateral support of structure,
- using 2"–4" thick foam panels for exterior sheathing (no skin),
- using single plate versus double plates for all walls,
- eliminating all headers in nonbearing walls and limiting them in bearing walls,
- using engineered wood for all main framing members for floors and roofs,
- connecting the walls at corners with lightweight clips, and
- removing all redundant framing members.

See Figures 4-17, 4-18, 4-19, and 4-20 for OVE framing details.
Optimum Value Engineering poses several hazards to firefighters.

- Cannot be readily identified during size-up
- Pushes the engineered design loads to the absolute maximum for framing members
- Strong as a complete system but extremely fragile when system is compromised by fire
- Load shifts will be more catastrophic as fire destroys members
- The foam panels significantly add to the fire load of the building

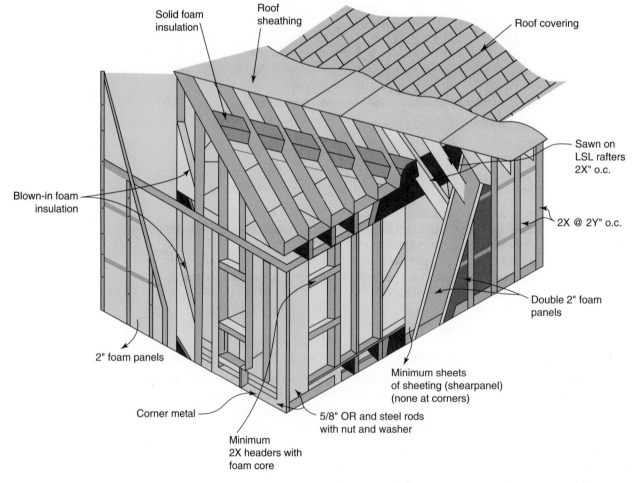

FIGURE 4-17 OVE framing details.

Shear panel (plan view)

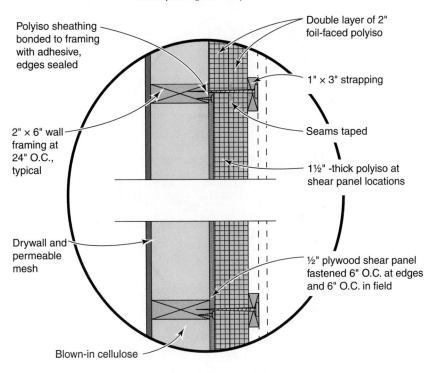

Polyiso sheathing bonded to framing with adhesive, edges sealed

Double layer of 2" foil-faced polyiso

1" × 3" strapping

Seams taped

2" × 6" wall framing at 24" O.C., typical

1½"-thick polyiso at shear panel locations

Drywall and permeable mesh

½" plywood shear panel fastened 6" O.C. at edges and 6" O.C. in field

Blown-in cellulose

FIGURE 4-18 OVE framing details.

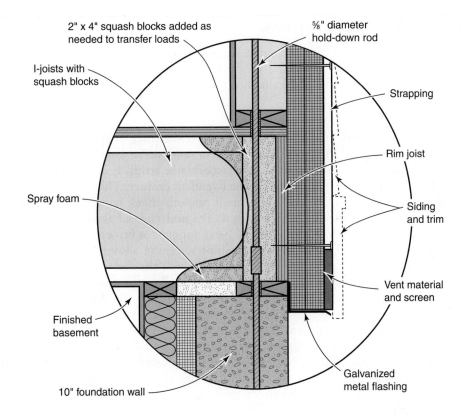

2" x 4" squash blocks added as needed to transfer loads

⅝" diameter hold-down rod

I-joists with squash blocks

Strapping

Rim joist

Spray foam

Siding and trim

Finished basement

Vent material and screen

10" foundation wall

Galvanized metal flashing

FIGURE 4-19 OVE framing details.

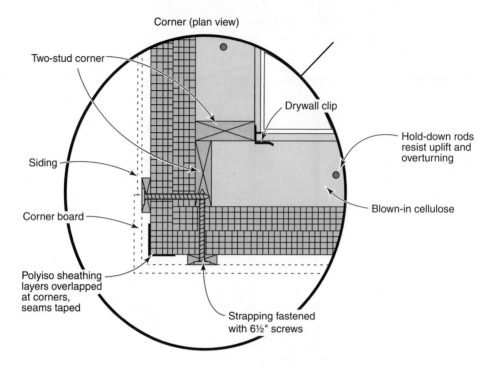

FIGURE 4-20 OVE framing details.

Corner (plan view)

Two-stud corner

Drywall clip

Hold-down rods resist uplift and overturning

Siding

Blown-in cellulose

Corner board

Polyiso sheathing layers overlapped at corners, seams taped

Strapping fastened with 6½" screws

Both of these systems are approved by the ICC for residential and some commercial applications. They will not be accepted by most codes in areas prone to high winds or earthquakes. The minimum fire code covering for these techniques is one layer of ½″ Sheetrock providing a fifteen minute rating.

Interior Wall Coverings

Interior wall coverings run a wide gamut. They include plaster, wood paneling, plastic, masonite, Sheetrock, or any combination of these. The wall coverings are attached to the studs by nails, screws, or glue or, again, a combination.

PLASTER

Plaster was predominantly used until the 1950s when Sheetrock or gypsum board became more practical to install. Sheetrock is discussed in more detail later in the chapter. The substrate for plaster was either wire mesh or wood lathe strips. It is not unusual to find lathe and plaster in buildings built before the twentieth century. This type of installation required several steps. The lathers, using small, smooth shank nails, installed the lathe strips horizontally over the vertical studs or to the underside of floor joists in random patterns encompassing up to three members. Next a rough, or brown, coat of plaster was installed using a trowel. The use of wire mesh or screening allowed for more coverage of an area with less craftsmanship. The brown coat was trowled directly onto the wire mesh. This coat often contained strands of horsehair to allow the product to stay together or bind better. Thus the nickname horsehair plaster evolved. The finish coat of plaster contained no horsehair and was almost pure gypsum. Additives were mixed in to give the drying time a boost, to prevent cracking, and for added durability. The problem with using plaster was the amount of moisture it contained. This moisture had to be allowed to dry fully before painting. This is evident in older homes that have blisters on the interior finish; blisters can also be the result of moisture leaking in from a faulty roof or plumbing fitting above. Elaborate shapes and patterns could be created with plaster, but all required the use of highly skilled craftsmen.

WOOD PANELING

Wood paneling and masonite are used as wall coverings to offset cost. The products are manufactured in 4′ × 8′ or 4′ × 9′ panels. The thickness of these panels is from ¼″ to 1″. The finish is transferred onto the panel to give the illusion of grained patterns and a variety of shades or colors. Textured wood panels were created to give more of an illusion of real wood. These panels, commonly known as T-111 or T1-11, are constructed of wood by-products but have grooves cut into them to add to the illusion of vertical wood panels. All these products are installed as stand-alones or can cover already existing wall coverings such as plaster or plasterboard. They are attached by nails, screws, or glue, or a combination of these.

Pre-finished panels with plastic, metal, or wallpaper are available. These types of panels still have a Sheetrock base and are commonly found in commercial installations rather than residences. Fire retardant materials with a fire-rated panel behind are required for installations involving life safety.

Many homeowners will install paneling over damaged plaster or Sheetrock wall coverings. This aesthetic cover-up adds to the fuel load of the room. The photocopied sheets are ¼″ thick and will greatly enhance the flame spread. Fire fighters should be prepared for these conditions at all times.

SECURITY WALLS

In many government buildings, buildings containing valuables, and in some residential structures there now exists a need to provide greater protection against terrorism and other man-created threats. For these reasons a firefighter may find walls constructed of steel, Kevlar, or bullet-resistant fiberglass. For aesthetic purposes the wall coverings may be Sheetrocked and many currently used tools and methods may not breach these walls. A solid knowledge of the building and thorough reading of building plans is paramount. Contact a local supplier or contractor to determine the correct tools and procedures to safely operate with these assemblies.

Roof

The **roof** of a building serves to protect the occupants from the elements and control the climate within. The earliest forms of coverings involved straw thatching. The straw was layered similarly to the husk around an ear of corn. The layers of straw shed water. This type of roof required a great deal of maintenance. Roof coverings evolve as progress in manufacturing occurs, but the principle for shedding water remains the same. A configuration of overlapping joints and shingle coverage has proven to be the most effective for watershed because gravity allows the liquid to run downward without wicking back up under the preceding layers. The use of bituminous products or tar led to their widespread use either as a binder for other materials, such as wood, or as a complete unit. Built-up roofs—layers of asphalt-impregnated paper with mopped-on layers of tar—are still being applied; however, rubber and neoprene membrane roofs are becoming more common because they require less maintenance and are of more consistent manufacture.

roof ■ The external covering of a house or building.

All modern roofs require the application of stone as the top coating. The stone acts to bind the asphalt as well as dissipates the heat generated by the roof's exposure to the sun. Slate and other stone products, as well as terra cotta tile, are quite pricey and require installation by craftsmen, so they are usually reserved for public buildings, commercial structures, or the homes of the wealthy. Slate look-alike products are being manufactured from resin and other lightweight materials.

The use of asphalt shingles has been common for the past fifty years, however, fiberglass products and plastics are overtaking them for most of residential usage (see Figure 4-21).

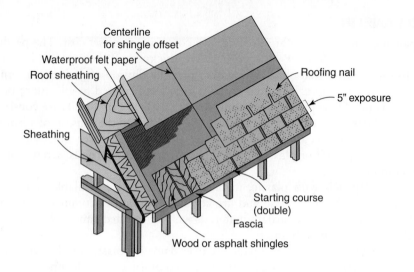

FIGURE 4-21 Installation of fiberglass or asphalt shingles.

Centerline for shingle offset

Waterproof felt paper

Roof sheathing

Sheathing

Roofing nail

5" exposure

Starting course (double)

Fascia

Wood or asphalt shingles

BUILDING GREEN

The use of recycled materials has led to a dramatic increase in the use of plastics and metal for roofing. Faux slate, cedar shakes, and Spanish mission (terra cotta) tile can be made from recycled plastics and are extremely realistic looking. Metal-standing seam roofing faded during World War II due to metal shortages. After the war, asphalt and fiberglass proved more economical. Now metal roofing products are almost exclusively manufactured from recycled materials. More training and procedures should be developed to operate safely on these systems.

Germany and other European cities have instituted the use of green roofs since the 1960s. The debate concerning global warming has increased the use of this assembly in the United States. Chicago, Illinois, Atlanta, Georgia, and Portland, Oregon as well as other U.S. cities are actively promoting the installation of green roofs.

Green roofs are defined as a roof system that is partially or completely covered with vegetation and which has been built on top of a waterproof membrane. Green roofs can contain ponds, which are used to treat gray water (see Chapter 5). Green roofs are beneficial to the environment for controlling rainwater, providing insulation to the building, offsetting urban pollution, reducing the heat island effect of urban cities, and providing wildlife habitats, thereby increasing the quality of life for the inhabitants of the building.

In many cities the runoff caused by rain is combined with the waste by-products of humans at water treatment plants. This process can, at times, severely tax the plants and can allow for more effluent (solid waste products) to enter the tributary system around the city. The use of a green roof will trap the rainwater and allow it to percolate through the soil medium and return to the atmosphere through evaporation and transpiration of the plants, thus reducing the amount of storm water runoff.

Another advantage of a green roof is the insulation benefits. For centuries, buildings in Europe have had green roofs for shedding water and providing insulation. It is thought that adding a layer of insulating dirt, rock, and other materials causes the building to require less heating and cooling while still maintaining optimum conditions for the inhabitants.

The green house effect, warming of the earth's surface by the emissions of methane, sulphur dioxide, and other by-products of carbon fuels, resulting in the destruction of the ozone layer around the earth and allowing the sun's rays to penetrate the atmosphere with greater intensity, is a matter of debate. It has, however, been postulated that as we tear down our forests to create urban environments, we destroy the earth's ability to produce vegetation-generating oxygen. The use of concrete, asphalt, and steel is prevalent in urban areas; these materials retain the heat of the sun and do not allow the heat energy to dissipate to the atmosphere at night, virtually acting as heat sinks. This phenomenon is called the heat island effect. The use of green roofs allows for the production of oxygen through photosynthesis, which helps to counteract the loss of vegetation that is a result of the creation of the urban environment. The vegetation on a green roof also assists with the liberation of heat energy by regulating the heat absorbed by the building.

Where the local codes will allow, grey water is captured after use in a holding pond on the roof where the water is treated and reused in toilets and as well as other building uses not requiring potable water. At the present time the *Uniform Plumbing Code,* adopted by many cities does not allow for the reuse of grey water.

The term *green roof* also pertains to roof systems that have photovoltaic cells (solar panels) installed on them. These will be discussed in Chapter 5.

The construction of rain roofs involves applying a membrane over the roofing substrate. A root barrier may also be installed so that vegetation does not penetrate the waterproofing membrane, which is usually made from rubber. Application of soil, peat, gravel, or other aggregates and vegetation; piping for irrigation and water recovery; retention ponds; and stone or concrete features for walkways complete the assembly. There are two types of green roof systems: extensive and intensive. An extensive system uses between 1″ and 6″ of soil and uses low profile plants, such as grasses, succulents, and mosses. Extensive green roof systems are becoming widely used. They weigh an average of 10 to 50 pounds per square foot and are generally only accessed for routine maintenance.

The intensive system uses a deeper soil level, in some cases over 2 feet. They are designed for crop propagation, shrubs, or trees. They weigh from 80 to 120 pounds per square foot. These are designed to allow the building's inhabitants to use the area for farming and relaxation. When a building is retrofitted with a green roof, complete engineering studies must be accomplished to ensure the roof will support the added load, especially in areas where heavy rain or snowfall is anticipated.

The residential market for green roofs is somewhat limited. Many residential roofs are pitched rather flat. Many local homeowner associations prohibit their application for aesthetic reasons. The use of mosses is common in residential green roofs; however, the use of solar panels or photovoltaic cells is rapidly increasing. These will be discussed in Chapter 5.

Green roofs pose several hazards to firefighters.

- Precludes the use of vertical ventilation in covered areas
- Fall and trip hazards exist due to vegetation and soil
- Roof may not have high parapet walls
- Any additional application of water from a hose stream may potentially cause collapse
- Can be found in all building types, but not all will be code compliant, nor will they necessarily have had engineering studies to determine the host building's ability to support them

All roofs need to have a slant or pitch to them. Even flat roofs are pitched (see Figure 4-22), allowing water to drain off. If we assume the weight of water to be 8.67 lbs per gallon, imagine the impact of pooling water on the structure. This is why areas that are exposed to high snow loads or rapid freezing suffer roof failures if improperly constructed. Sometimes this occurs from something as simple as a blocked scupper. *Scuppers* are openings in floors or roof assemblies that allow water to be removed.

(a) Shed roof

(b) Gable roof

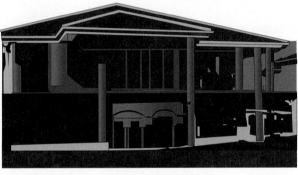

(c) Hip roof

(d) Gambrel roof

(e) Dutch hip roof

(f) Modified mansard roof

(g) Flat roof

(h) Intersecting roof

FIGURE 4-22 (*continued*)

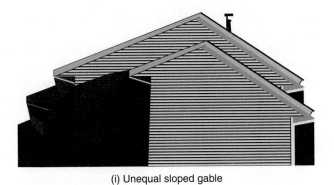

(i) Unequal sloped gable

FIGURE 4-22 Various roof configurations: (a) shed roof; (b) gable roof; (c) hip roof; (d) gambrel roof; (e) Dutch hip roof; (f) modified mansard roof; (g) flat roof; (h) intersecting roof; (i) unequal sloped gable.

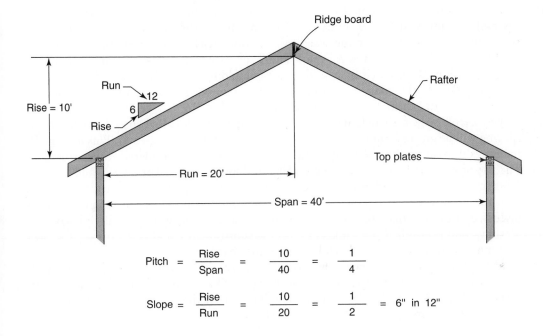

FIGURE 4-23 Calculation of roof pitch.

$$\text{Pitch} = \frac{\text{Rise}}{\text{Span}} = \frac{10}{40} = \frac{1}{4}$$

$$\text{Slope} = \frac{\text{Rise}}{\text{Run}} = \frac{10}{20} = \frac{1}{2} = 6'' \text{ in } 12''$$

The main structural members of a roof are called *rafters* (in wood construction) and *joists* or girders (in steel assemblies). The pitch or slant of a roof is expressed as a fraction, for example, $4/12$. This fraction tells us the roof rises 4″ in height for every foot of span. The span of a roof is figured by measuring the width and dividing it in half. The height of a roof is determined by multiplying the rise (in inches) by the span. For example, a building that is 30′ wide would have a 15′ span. Using our earlier example of a 4″ pitch, the height of the roof would be 180″ (15′ × 12″ × 4″ = 720. Divide this number by 12 and the height is 60″ or 5′.). This measure will be the vertical height above the top plates of the upper story. A pitch of up to $6/12$ is safe and comfortable to work on. Any pitch above $6/12$ is considered unsafe because the steepness will not allow you to remain upright and perform work. These roofs are best handled from an aerial ladder, platform, or tower (see Figure 4-23).

Wooden roof rafters are usually 2 × 8, 10, or 12. The span and pitch also dictate what type of roof covering can be applied to a sized rafter. For example, rolled roofing is not permitted by many codes for pitches under $4/12$. Slate and terra cotta, heavyweight roof coverings, require pitches above $6/12$ and rafter sizes of 2 × 12. *Note:* Imitation slate and terra cotta, which resemble actual slate and terra cotta, use lighter-weight materials.

Roof coverings for commercial structures will run the gamut of materials depending on the age of the structure and its original purpose.

Modern commercial buildings are most often required to be of Type I or II construction. This should lead us to assume that the roofing materials will either be built-up, rolled rubber membrane, or neoprene over insulation with concrete decking (which may be on Q decking). Steel bar joists with steel main girders should support the assembly; however, if the building was originally intended as a residence, or was built before the code changes, then the roof could very well be constructed of wood rafters and decking. It might even have large trusses with either cast connectors or split-ring connectors. Most high-rise buildings will either have built-up or membrane roofs.

Floors

floors ▪ In architecture and building, the distinct stories of a structure are referred to as floors.

Floors form the horizontal platforms within a building. The main components of a floor assembly are joists, substrate, underlayment, and the finished floor. The joists form the supporting platform to carry the weight of the floor itself, the furnishings, and the occupants. In conventional framed Type V and Type III structures the size and spacing for the joists will be determined by a section of the building code. The size and spacing will be determined by the span (the distance between supports holding the joists) and the anticipated load. Normally, the spacing will be 12" or 16" on center. In general, the longest usable span for single wood joists is about 15" to 20'. In Type IV structures the spacing will be greater, but the size of the joists will also be much greater. Plus in Type IV buildings the use of 3" to 6" tongue-and-groove planking is considered to be structural. The use of trusses allows for greater spacing and longer spans. For Type I and II structures the substrate will either be concrete alone or concrete poured over metal Q decking with rebar and wire mesh. Substrate refers to the component that forms the platform. As discussed, it can be concrete, metal, or wood. The underlayment is normally plywood or masonite and is installed perpendicularly to the substrate in wooden buildings. It further strengthens the assembly and provides a smoother base.

Most floor assembly supports are in compression at the ends of the joists, but in Type V balloon–framed buildings it is very often in shear. The finished floor coverings can be carpet, tile, wood, marble, slate, or any combination of these. The only limitations are weight and economics.

Ceilings

ceilings ▪ The overhead surface of rooms.

Ceilings are the coverings of the underside of the floor joists above. They can be attached directly to the joists or dropped. There are many different forms of ceiling materials. The most common in residential construction today is Sheetrock.

Sheetrock goes by many names: wallboard, plasterboard, gypsum, or gypboard. Sheetrock consists of gypsum. This mineral is found all over the world. In its purest form it is white, but impurities such as clay and iron oxides can render it gray, brown, or even pink. Once it is mined from the earth it is baked to 350°F and crushed into small particles. This fine powder is then mixed with aggregates, additives, and fibers to give it moisture resistance and added strength. Water is then added and the mix is sandwiched between layers of paper. There is a finished (or smooth) side and a rough side. Sheetrock replaced plaster as a wall and ceiling covering primarily because it was dry when installed (thus another nickname: *drywall*). The panels or sheets of Sheetrock come in 8', 10', 12', and 16' lengths and almost always in 4' width. The thickness of the product varies also. It comes in 1/4", 3/8", and 1/2" thicknesses for regular assemblies. It also comes in 5/8" and 1" thicknesses when there is a need for fire stopping.

Sheetrock possesses *endothermic* properties, meaning it will absorb heat rather than allow heat to pass through it. By its nature and composition, Sheetrock, when heated, liberates hydrates contained within its molecular structure. Sheetrock sacrifices its structure, which prevents fire spread. It is referred to as passive fire protection.

In general, each layer of $1/2''$ Sheetrock yields fifteen minutes per hour of fire-stopping ability. Two layers of $5/8''$ Type X rated Sheetrock will yield one hour of fire resistance. Type X is different in that it contains glass fibers, perlite, vermiculite, and boric acid in order to increase its fire resistance.

SUSPENDED CEILINGS

The dropping of ceilings is a common occurrence in commercial assemblies and is often found where basement or room remodeling has taken place. The advantages of dropping a ceiling are increased sound proofing and better access to utilities above. The area between the bottom of the joists and the top of the panels is called the *plenum*. This space is often used by HVAC tradespeople to move the air back to the air-handling unit. This then becomes the return air duct. Dropped ceiling panels come in either $2' \times 2'$ or $2' \times 4'$ sizes. A lightweight grid system is installed in tension to the undersides of joists or roof rafters. This assembly can have specially manufactured panels and stronger grid parts and be rated fireproof, but often it is not.

The problem for firefighters is that the grids are suspended by thin wires, which fail rapidly when exposed to temperatures as low as 600°F.

The earliest version of dropped ceilings was pressed tin. These panels were tongue-and-groove, thus allowing for a seamless system. The panels were attached to strips of wood applied perpendicularly to the run of the joists. These types of units are also called *purlins*. Tongue-and-groove simply means there is a male component that fits into a pocket, thus sealing the joint. This same system is employed with wooden plywood underlayment and also with plank flooring. Some of the most difficult ceilings for a firefighter to encounter are pressed tin and plaster over wire mesh. The problem with these assemblies is that traditional pulling tools, such as hooks or pike poles, are limited in separating the joints or cutting the mesh. They will be found in many late nineteenth and early twentieth century buildings, even in buildings that have been renovated.

Doors and Windows

Doors allow occupants to move in and out of a structure while allowing the premises to be secured when desired. Windows or glazing allow the occupants to see outside while still being protected from the elements.

Doors come in a wide array of installations, styles, and materials (see Figure 4-24). Normally, door assemblies consist of a door, jamb with hinges, and handle or lock set. Doors will either be hollow or solid core. The swing of a door is the direction the door travels when it is opened. This can be determined by putting your back against the hinge side of the jamb; whatever direction (either left or right) the door moves is the swing (see Figure 4-25).

The size of a door is determined by measuring the slab or door itself, not the assembly. So a $3' \ 0'' \times 6' \ 8''$ door would actually be $3'$ wide by $6' \ 8''$ tall. The jambs can either be split or solid. They can also be made of either wood or metal. The use of split jambs involves two sections, which encompass the jacks and header of an opening using a tongue-and-groove method. These are widely used in residential construction and have replaced solid wooden jambs. Most residential exterior door assemblies consist of a solid-core door with a solid jamb, thus allowing for increased security. All wooden jamb assemblies are either nailed or screwed into the framing members.

In commercial installations metal jams are more widely used. These can either be knock-down or solid jambs. The knock-down assemblies come in three sections—left, right, and the head. They are fitted together inside of a framed opening and attached to the wood or metal framing members. The wall covering is then inserted into a space

FIGURE 4-24 Symbols on plan view for door configuration. *Source: Dagostino, Frank R.; Feigenbaum, Leslie, Estimating in Building Construction, 6th Edition, © 2003, p. 269. Reprinted by permission of Pearson Education, Inc., Upper Saddle River, NJ.*

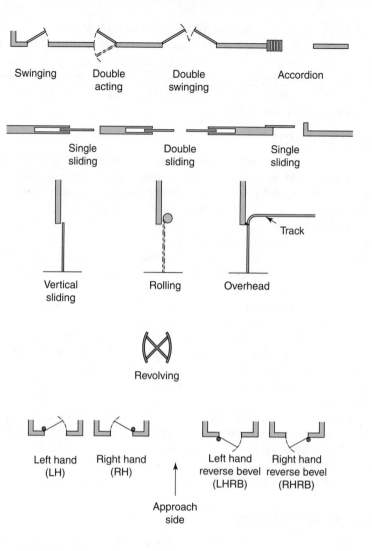

Swinging   Double acting   Double swinging   Accordion

Single sliding   Double sliding   Single sliding

Vertical sliding   Rolling   Overhead   Track

Revolving

FIGURE 4-25 Door hand and swing. *Source: Dagostino, Frank R.; Feigenbaum, Leslie, Estimating in Building Construction, 6th Edition, © 2003, p. 269. Reprinted by permission of Pearson Education, Inc., Upper Saddle River, NJ.*

Left hand (LH)   Right hand (RH)   Left hand reverse bevel (LHRB)   Right hand reverse bevel (RHRB)

Approach side

between the jamb and the framing member, forming the seal. After installation, these assemblies are quite strong. Solid metal jambs contain the left, right, and head sections as a unit and are usually installed as construction takes place. They are attached in the same manner as knock-down jambs. Most often these assemblies will use either solid-core or metal door slabs. The swing for these assemblies can be figured in the same fashion as wooden assemblies.

A hollow-core door assembly has solid wood or metal around its perimeter but contains a corrugated fill within the panel. The skin of the door is then attached to the solid members. The skin of these doors can be metal, plywood, plastic, or masonite. A solid-core door is just that. It will not contain the cardboard fill. Paneled doors are considered to be solid core but they can be constructed of either wood or masonite. Doors can contain glass (or *glazing,* which is the term used in the construction industry) but that does not change the terminology hollow or solid.

There are many types of security systems for doors. Basically, the more security needed the better the lock system will be. Sometimes you can defeat the security by going through the building system rather than the lock system. For example, in one large hotel fire firefighters in a stairwell found a locked out-swing metal door in a masonry wall. Armed with only an axe, sledgehammer, and halligan bar they took out one concrete block and reached in to activate the panic bar. They were able to get in and made sixteen rescues that night. The entry took less than two minutes.

WINDOWS

Windows are the eyes of a structure. The discovery that sand, which is also called silica, could be heated into liquid form and then cooled to produce a material that is see-through allowed people to finally seal out the elements. The Roman Empire is given credit for this discovery, but it was not until the Middle Ages in Europe that the use of glass proliferated. Stained or painted glass was widely used in castles and churches. Interestingly, this process originated to cover up defects—air bubbles and glass distortion in the panes—in early glass production. The form of glass that we know today came about at the end of the seventeenth century with the advent of the Perrot process, which places the molten ingredients into a mold and then spreads the mixture flat with rollers.

The raw material for making glass is sand mined from sandstone deposits. The major ingredients added to sand are sodium oxide and calcium oxide, which is why flat glass is referred to as soda-lime glass. Iron oxide is also present in raw silica and is what gives glass its bluish tint when viewed from the edge. The sodium oxide acts as a flux, thereby reducing the impurities and lowering the melting temperature of the silica. Calcium oxide stabilizes the mixture of molten glass and allows for the gas bubbles to be released; it also makes the mixture easier to handle.

Most glazing used in today's construction is known as annealed glass. Annealed glass is derived by floating the molten mixture of silica in a tin bath (see Figure 4-26). Tin has a higher specific gravity than molten silica; therefore the glass floats. This glass, however, breaks into shards when breached.

For safety, and to provide firefighters access to high-rise building window systems, the use of tempered glass is recommended. This process involves melting the silica mixture and cooling it rapidly by blowing jets of cooled air over it. This allows the exterior of the product to harden but not the interior and creates a dicing effect when the glass is broken. *Dicing*, the breaking up of the glass into thousands of small, relatively nonsharp pieces, provides the safety factor (see Figure 4-27).

Glass can be tinted by adding metallic pigments to the batch. To create reflective glass, metal or metal oxides are applied when the glass is still in its molten state but is in the process of being cooled. This process is called pyrolytic deposition. Chrome, stainless steel, titanium, gold, and copper oxides are used for this purpose (see Figure 4-28).

The building industry is developing better systems designed to provide greater insulating qualities or prevent heat gain from the sun. One of these systems is low-e glass in which a thin sheet is installed in the glass mixture. This film prevents the negative effects of the sun's rays as well as preventing heat from escaping in colder climates. Insulating glass is also used. In this assembly the panes of glass are doubled and separated by a closed channel. This channel is filled with desiccant granules or gas (commonly krypton or argon). This process absorbs moisture generated by the heating and cooling of the assembly.

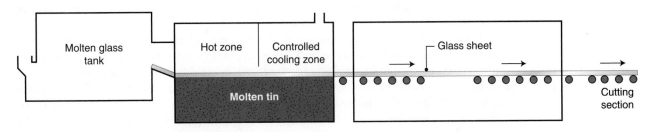

FIGURE 4-26 Outline of float glass manufacturing process.

FIGURE 4-27 Breakage
patterns of tempered
and annealed glass.

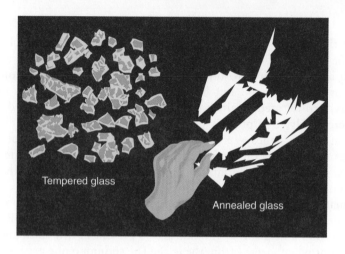

The structured assembly of windows also comes in many forms. Individual panes are called lites. Thus a six-lite window has six panes of glass per section (or sash) (see Figure 4-29).

A window system that cranks out at 90 degrees is referred to as a casement window. The types of windows that raise and lower are referred to as single- or double-hung. Only one panel moves in single-hung windows; in double-hung, both move. Windows that contain horizontal lites, either glass or acrylic, or wooden louvers, and open similar to an awning, are called jalousie windows. Jalousie windows with lites wider than 6″ are called awning windows.

Window assemblies can be constructed of wood, metal, plastic, or a combination of these materials.

FIGURE 4-29 Metal frame sash windows. *Source: Peter Bush © Dorling Kindersley.*

Stairs

Stairs transport occupants from one level of a building to another. Stair components are stringers, risers, treads, and the rail assembly (see Figure 4-30). Stringers are the members that carry the whole assembly. Treads are the surfaces the occupants walk on. Risers are the vertical faces of the treads. The rail assembly can be quite elaborate. In older homes the banister and spindles are mortised into the assembly, while in newer homes they might not be. This can be hazardous for firefighters who are dragging lines and passing one another on the stairs. Any members not mortised will fail quickly and catastrophically, possibly sending firefighters down to the floor below.

Approximately 40 square feet are needed to create an opening that will allow sufficient headroom for occupants to go from one level to another. This equates to an opening

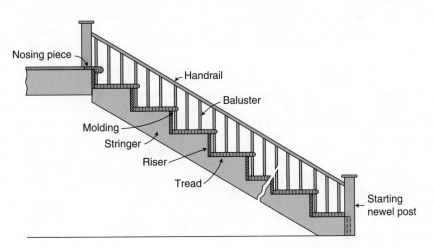

FIGURE 4-30 Parts of a stair.

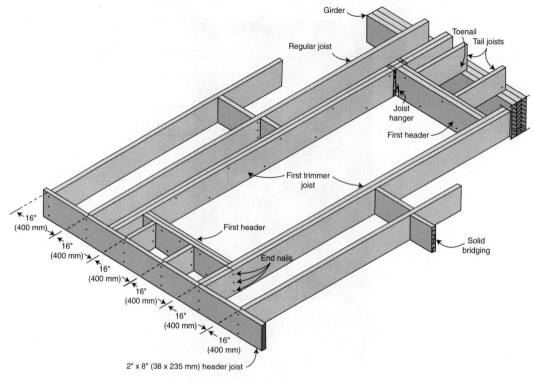

FIGURE 4-31 Stairwell opening with wood framing.

approximately 4' wide and 10' long for an elevation of 8'. The framing members are doubled around the hole so that the opening can be structural (see Figure 4-31).

The stairs themselves are not part of the structure's integrity. They carry only themselves and the weight of the occupants (see Figure 4-32). Therefore in multilevel structures the stairs are stacked one above the other. Or, if there is a basement, the interior basement stairs are directly under the stairs going to the second floor.

SCHOOL OF HARD KNOCKS

I spent the bulk of my career prior to becoming a chief working in truck companies. To perform the activities associated with truck work required knowledge of tools and knowledge of building construction materials and methods. My years of experience in these areas allowed me to become a better ladder man and carpenter. I was able to use more technique and less brute force. And I became able to perform these tasks much more efficiently.

As a chief officer I needed to be able to estimate how long the activities would take and how many firefighters would be required to complete the tasks safely and effectively. This is critical, in my opinion, so that firefighters are not exposed to hazardous conditions for longer than is necessary.

I put my knowledge and experience to use when we were attacking a fire in the roof area of the U.S. Treasury Building. The area affected was 40' wide by over 200' long. The ceilings were comprised of two layers of plaster over wire mesh. These ceilings were 14' high. I knew from experience that breaking through these ceilings would be an extremely difficult operation to perform. So instead of employing large numbers of truckies with hooks, I ordered the pumping pressure raised into the standpipe system and by using 2½" handlines with straight tips we literally blew the plaster and mesh from the rafters and joists. One firefighter can man a large caliber stream. Loop the line and have the firefighter sit on the loop. To ventilate, I had the units raise the windward windows and close the leeward windows; this pressurized the space. But remember: the wind direction and velocity will be different at higher floors than at ground level.

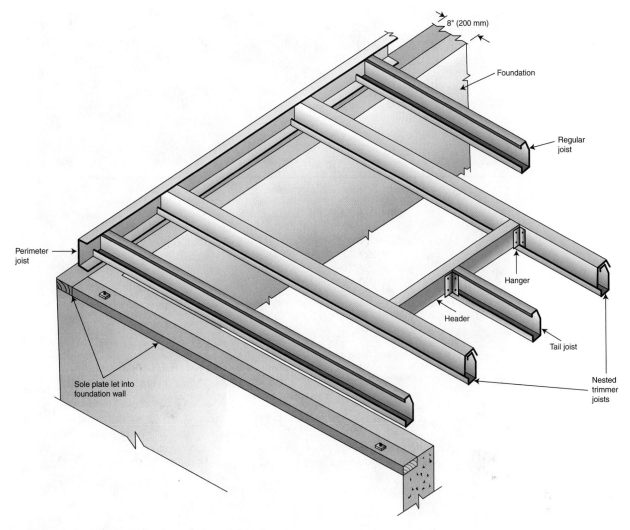

FIGURE 4-32 Framed opening for stairs in steel floor frame.

Units on the scene are responding to a MAYDAY call. The initial team has reached the victim, a firefighter. They report that the victim is located on the second floor of a four-story building. The victim has suffered a blow to the head and neck due to a ceiling collapse involving plaster and lathe and is unconscious. The rescue crew wants to bring the firefighter out quickly. The first floor is brick but the upper floors are covered with aluminum siding. There are many 4′ wide windows on each floor. The stairwell is heavily congested with hoselines and active firefighting is still going on. The fire is located on the second floor and involves five rooms.

1. As the incident commander, what would be your first concerns regarding elapsed time of operations?
2. As a truck company officer, what tools would you need to create an opening and where would you make it?
3. Could you use your aerial ladder or tower ladder to crane the victim down and what would you need to do so?

A myriad of materials, from marble to carpet, can be used to finish treads and risers. Stringers can be metal or wood. Stairwells can be open or closed. If one side is open to below, then they are called open. If both sides are walled, then they are called closed. The principles for spiral stairs are the same.

Summary

Foundations form the base of a structure and floors transmit the loads along a horizontal plane. The exterior walls form the vertical supports to transmit all loads from the roof back to the ground. Many terms are used to describe walls. Understanding how each type of wall works will assist fireground activities. Understanding door and window construction will allow firefighters to manipulate these systems more effectively and safely. The advent of green building is posing some unique challenges and dangers to firefighters.

Review Questions

1. Discuss how windows and doors are installed in a structure.
2. Discuss roof pitch, span, and rise and why they are important.
3. What is the maximum safe pitch?
4. Discuss the ramifications of the following during rapid intervention:
   - Doors
   - Windows
   - Stairs

5. Discuss the problems with thermal pane glass as it relates to size-up and ventilation.

Suggested Readings

Dunn, V. 1988. *Collapse of Burning Buildings*. Saddlebrook, NJ: PennWell Books.

Smith, J. 2002. *Strategic and Tactical Considerations on the Fireground*. Upper Saddle River, NJ: Pearson Education.

Terpak, M. 2002. *Fireground Size-Up*. Saddlebrook, NJ: PennWell Books.

Reimer, M., Roland, A., Wolfgang, A., 2009. *Green Roofs—Bringing Nature Back to Town*. International Green Roof Association.

Building Systems

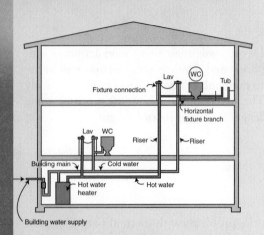

Fixture connection
Lav
WC
Tub
Horizontal fixture branch
Lav WC
Riser
Riser
Building main
Cold water
Hot water heater
Hot water
Building water supply

OBJECTIVES

After reading this chapter, you should be able to:

- Understand the building systems contained in residential and commercial structures.
- Define important terms used by tradespeople.
- Better understand how to mitigate problems properly utilizing the building systems.

Resource Central

For additional review and practice tests, visit **www.bradybooks.com** and click on Resource Central to access book-specific resources for this text! To access Resource Central, follow directions on the Student Access Card provided with this text. If there is no card, go to **www.bradybooks.com** and follow the Resource Central link to Buy Access from there.

Introduction

In this chapter we will discuss the various systems of the structure. It is important to understand how buildings are constructed and how the subsystems are installed. If we understand the relationships between all the components we can make better decisions regarding size-up or overhaul. We also need to comprehend the radical difference in size and scope between residential and commercial systems. During emergencies in commercial buildings we need to communicate with utility company personnel and the building's staff; therefore we need to use their language to get the job done.

FIREHOUSE DISCUSSION

The MGM Grand incident occurred on November 21, 1980, killing eighty-four civilians and injuring an additional 679 civilians. The fire also injured fourteen firefighters, most from smoke inhalation.

The MGM Grand was a twenty-six story T-shaped hotel/casino complex. The building measured 380' × 1200' at ground level. The casino alone was 150' × 400'. The remaining space at ground level was occupied by restaurants, showrooms, and a convention center as well as the upper portion of a jai alai fronton. The part of the building below this portion of the building was approximately the same dimensions as the upper level and contained the arcade, the lower portion of the jai alai fronton, a movie theater, a large number of shops and boutiques, service areas, and underground parking.

The hotel portion consisted of three T-shaped wings each measuring 320' long and 75' wide. The hotel portion contained 2,076 guest rooms. At the time of the fire, there were approximately 5,000 guests, staff, and other people in the building.

The building was of mixed construction. The construction types included fire-resistive, protected noncombustible, and unprotected noncombustible. The interior finish varied significantly and included both combustible and noncombustible materials.

The building complex was partially sprinklered. Protected areas included the arcade level, major portions of the casino level, and part of the twenty-sixth floor. The code at the time of construction called for sprinklers to be installed in any public areas larger than 12,000 square feet. It was interpreted, however, that the restaurant and casino areas would be functioning twenty-four hours per day and thus any fire would be discovered. This interpretation led to the areas involved not to be sprinklered.

The means of egress from the casino level was either through doors directly to grade or down one flight to grade. The egress for the high-rise portions consisted of one stair and one **smoke-proof tower**, a stairway providing direct access to outdoor air at each floor level or pressurized to prevent inflow of smoke into the area. All stairs discharged to the exterior of the building at ground level. All of the stairs and the one smoke tower were not enclosed with two-hour rated construction. For security purposes, once you entered a stairway there was no reentry to other floors.

There were four major subsystems for the heating, ventilation, and air conditioning (HVAC) system. Heated or cooled air was supplied through ducts for the arcade and casino levels. Air returned through transfer grills, open grates, and the lighting system to a large return-air plenum above the ceiling. This is done in most high-rise buildings and most large area box stores. The return-area plenum for almost the entire casino was through one undivided area. In most properly designed systems, the use of smoke curtains or barriers will limit smoke spread upon activation of the alarm.

The second subsystem provided conditioned air from a mechanical *penthouse*, which is any structure above the roof line, to the three wings of the high rise and the *central core*, which is the area that contains the elevators, public bathrooms, electrical, and plumbing runs as well as any other major system supporting the building. This central core area remains constant throughout all vertical levels independent of the use for that space. For example, in office buildings where tenants will move in and out as leases commence or expire, the tenant can reconfigure the floor space usage except for the core area, which will be contained within demising (also spelled *dimising*) walls. This core can be located in the center of the building thus the term center core, but it can also be on either the left or right side and correspondingly called right or left core. These walls are normally built to a three to four hour fire rating, but check with your local codes for the rating in your area. The term may also be used to indicate walls between tenant spaces, although the trades in that context rarely use it. The penthouse also housed the elevator machinery for one bank of elevators.

There was no return air system for the tenant corridors. Air movement was created by providing makeup air to the guest rooms, which used individual fan units with chilled water pumped to them.

smoke-proof tower ■
A stairway that provides direct access to outside air at each floor level or is pressurized to prevent inflow of smoke into the area.

Makeup air allows for air to be vented to the outside and a source then provides replacement air via vents or fans. The guest rooms constituted the third major subsystem

The fourth major subsystem was the exhaust fan ducting for the toilet exhaust for all floors. The area around this rooftop duct outlet is not a pleasant spot to linger for long. There were also two seismic joints that went from the area above the ceiling of the casino to the return-air plenum above the twenty-sixth floor. *Seismic joints* are continuous gaps between two or more adjacent and connected structures allowing them in the event of an earthquake, to move back and forth without colliding. The shafts were approximately 1' wide. These joints isolated the east and west wings from the south wing. The bottoms of the shafts were not enclosed and opened directly to the plenum area above the casino. There were flexible, nonrated, accordion-fold stainless steel panels floor to ceiling on each level in the walls where the corridors crossed these seismic joints.

A manual fire alarm system with bells and public address capability was provided in the building. There did not appear to be manual pull stations on the arcade or casino levels (remember it was thought that these areas would be staffed and in use twenty-four hours a day); the system could be activated at a security station in the casino. The guest room floors were equipped with manual pull stations. This was a local signaling system only; it did not notify the fire department. This type of system is a **pre-signal system**. Once an alarm is sounded, the signal goes to a manned area on-site. This could be the front desk, security office, building engineer's office, or some other constantly occupied location. It then requires staff to physically investigate the problem and, if confirmed, notify the fire department. This was the same system in place at the Meridian Plaza Fire in Philadelphia in 1991. If there is not a system for off-premises reporting that automatically notifies the fire department, then the entire system relies on human response leading to at least delays in reporting.

The fire began in a restaurant on the casino level. It is believed that an improperly grounded wire subjected to vibration eventually shorted out and the fire began. The fire burned without detection for some time. Early attempts by building staff to put out the fire had no effect and the staff was forced to retreat. The fire grew significantly, feeding on combustibles within the restaurant. The smoke and heat entered the plenum area above the casino. The restaurant area flashed over and provided the thermal energy to begin affecting the casino. Once again, large amounts of flammable materials, foam padding, and moldings contributed to rapid flame spread. Soon the casino area flashed over.

Units from the Clark County Fire department responded within two minutes of the receipt of the call to the fire department. The scene that faced them was thick black smoke emanating from the casino area and a large fireball erupting from the restaurant area. Very little smoke was evident on the exterior of the building at this time. In a short time the casino erupted in solid flame. It was estimated that the flame spread in the casino was traveling at a rate of 15' to 19' per second. Other units soon joined the first units, eventually involving 554 firefighters from Clark County and mutual aid departments. Through heroic efforts and aggressive interior fire attack the fire was quickly knocked down, however, the damage was done.

The smoke and heat had been picked up by the plenum area and transferred to the seismic joints as well as other ducts, ducts that had improperly-installed smoke dampers that prevented them from closing. In some cases fusible links were improperly installed and in others the dampers were improperly connected and the bolts prevented them from closing. The locked doors in the stairways trapped many of the victims who died in place. Some died in the casino while others died in their rooms. This incident led to Nevada developing some of the most stringent codes in the country.

pre-signal system ■ A signal sent to a constantly occupied, manned area on site after an alarm sounds that requires staff to physically investigate the problem and to notify the fire department once confirmed.

Residential

After a structure has been framed in or enclosed various systems are installed to provide climate control, power, and light and to distribute water and remove waste.

PLUMBING

The plumbing system is probably the most valuable to the occupant. You can do without many things, but you can't do without water. Since the earliest forms of plumbing, gravity is the propelling force behind water delivery and waste removal. There are two completely separate systems for the delivery of fresh water and the removal of waste. A system of valves can control flow from a city's water supply through the underground network of piping to the meter for each structure serviced by the system. After the meter measures usage, control is handed over to the occupant via a set of smaller valves located at each

THINK ABOUT IT!

Many firefighters are unaware of system shortcomings for buildings in their jurisdiction. When firefighters assume that a building is totally sprinklered because it has a Siamese connection, they might be dead wrong. The alarm system may be similar to the one installed in the MGM Grand. All departments must conduct familiarization tours of all multiple-occupancy structures to answer these questions as well as many more. All high-occupancy facilities such as bars, apartments, churches, and high-rises must be sprinklered or fatalities will occur from smoke inhalation.

- Look at your local code and ascertain if all buildings with tenants or guests have to be 100 percent sprinklered.
- Discuss your priorities upon arrival at an incident like this if you have limited personnel.
- What would you include in a pre-plan for this building?
- Discuss the chain of custody for evidence (including bodies).
- Discuss what size hoselines should be deployed by the first units for a fire of this magnitude.
- Discuss what methods are necessary when confronted with a large panicked crowd trying to exit and what can be done to control the crowd while considering firefighter safety.

appliance. Appliances come in many forms and materials. For buildings utilizing a well, the system involves a submersible pump, piping to a pressure tank (within the structure), and connections to the building's piping. See Figure 5-1 for a typical installation of a well system and Figure 5-2 for a pressure tank.

Bathroom areas pose significant problems for firefighters because of the use of one-piece fiberglass tub enclosures and the use of tile on the walls. Both can be hindrances if the firefighter is attempting to evacuate an adjoining room by way of a wall breach. Proper size-up and risk management coupled with pre-planning can help to prevent this scenario. Upon arrival, if the roof is visible, look for the vent stack sticking up through the roof covering. The bathrooms and kitchen(s) will be within 5′ in either direction longitudinally from this stack (see Figure 5-3). Firefighters should make it a priority to search walls adjoining bathrooms and kitchens first when searching a room.

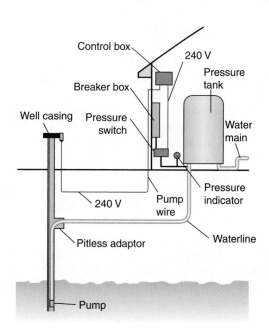

FIGURE 5-1 Typical installation of residential well system.

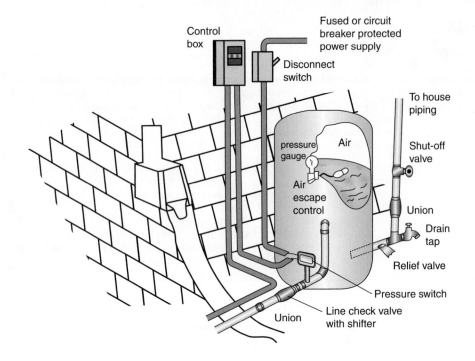

FIGURE 5-2 Pressure tank for residential well system.

Control box

Fused or circuit breaker protected power supply

Disconnect switch

To house piping

pressure gauge

Air

Shut-off valve

Air escape control

Union

Drain tap

Relief valve

Pressure switch

Line check valve with shifter

Union

The major components of a residential system for urban and suburban occupants are the meter, hot water heater, the piping, and the appliances (see Figure 5-4). The hot water heater can be powered by electricity or natural gas or can be part of a boiler system that uses fuel oil. The principle is simple: the water is retained and heated to a preset temperature and then delivered under low pressure to the appliances throughout the structure. As the water is used it is replaced by fresh water and the process begins again (see Figure 5-5).

Because heat develops pressure within an enclosed vessel, there is a safety valve mounted on the water tank. This valve will open if a certain pressure is reached. This valve should never be tampered with; any evidence of tampering and the system should be made inoperative. Valves and pipe sizes are used to control the pressure and volume delivered.

The most common materials used today for the conveyance of water are copper and polyvinyl chloride (PVC). The use of PVC is gaining acceptance in plumbing codes. PVC and acrylonite butadiene styrene (ABS) are widely used as vent and drain systems. These have replaced cast iron. Chlorinated polyvinyl chloride (CPVC) is gaining acceptance for use as supply, vent, or drainpipes. It is as easy to use as PVC and as reliable as copper. Figure 5-6 lists materials used in residential plumbing systems.

The Uniform Plumbing Code is a national set of codes that is updated every three years; local codes can vary from this set of codes and often do by including additional restrictions. Earlier forms of galvanized pipe are no longer used. The joints of a copper pipe system use a flux to clean impurities and solder. Older homes will most likely have a 50/50 tin/lead solder, allowing lead to be introduced into the water supply, which is a cause for health concerns. Modern solder is lead-free. Most supply lines are ¾″ copper, which steps down to ½″ at the appliance shutoff valve. Flexible copper or plastic lines are widely used from the valve to the faucets or bibs. The flex lines use compression or flared fittings instead of solder. The type of copper pipe in most residential uses is known as type M. It is the thinnest and cheapest copper pipe. Type L is used for commercial applications; type K is the thickest and is never used for residences.

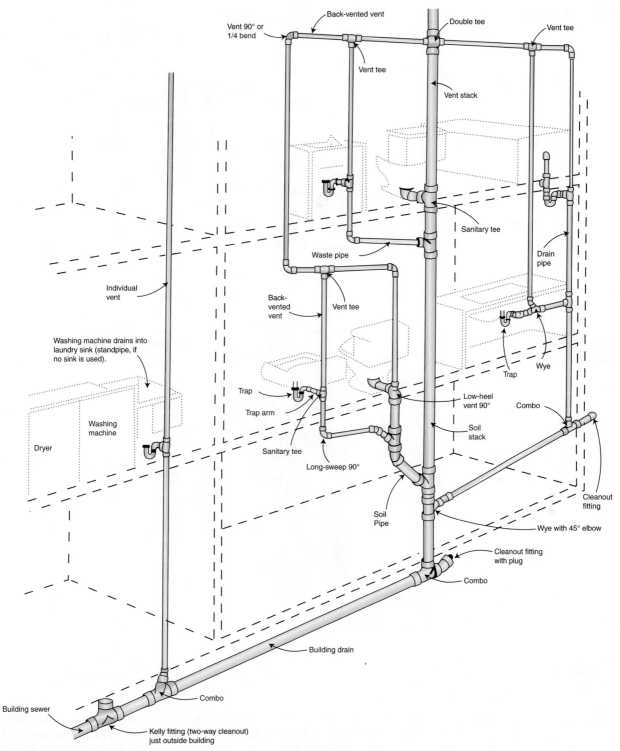

FIGURE 5-3 Stack through roof will identify locations of fixtures in buildings.

Labels in figure:

Vent 90° or 1/4 bend
Back-vented vent
Double tee
Vent tee
Vent tee
Vent stack
Individual vent
Sanitary tee
Waste pipe
Drain pipe
Back-vented vent
Vent tee
Washing machine drains into laundry sink (standpipe, if no sink is used).
Trap
Wye
Trap
Low-heel vent 90°
Combo
Trap
Trap arm
Soil stack
Washing machine
Cleanout fitting
Dryer
Sanitary tee
Long-sweep 90°
Soil Pipe
Wye with 45° elbow
Cleanout fitting with plug
Building drain
Combo
Combo
Building sewer
Kelly fitting (two-way cleanout) just outside building

FIGURE 5-4 Typical plumbing system.

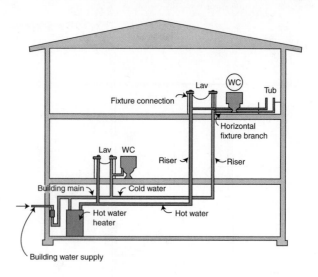

FIGURE 5-5 Residential water heater. *Courtesy: State Industries, Inc., Ashland, TN.*

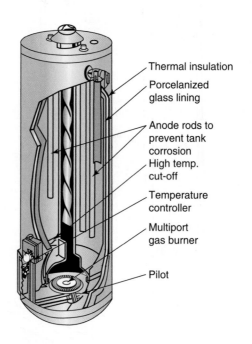

- Thermal insulation
- Porcelanized glass lining
- Anode rods to prevent tank corrosion
- High temp. cut-off
- Temperature controller
- Multiport gas burner
- Pilot

blackwater ■ Waste water containing effluent.

gray water ■ Water from vanities, washing machines, dishwashers, and showers.

fall ■ Term used by the plumbing industry to define rate of declination for lines carrying effluent.

DRAIN SYSTEM

Drain systems in the residential sector are fairly straightforward. Unwanted waste products, **blackwater** (waste from toilets) and **gray water** (from showers, bathtubs, and washing machines) are delivered to an ABS pipe or stack to be delivered to a septic system or the urban waste treatment plant. It is still the rule of thumb that drainpipes should have a ¼″ per foot **fall** to facilitate the movement of both liquids and solids through the pipe. More fall and the liquid would travel too fast to push the waste and paper materials. Less fall and the liquid would travel too slowly and the mass or effluent would begin to settle after the charge of water has passed by. The P traps under the appliances are shaped to

FIGURE 5-6 Materials used in residential plumbing systems.

MATERIALS FOR POTABLE WATER (1) (2) (3)	WATER SERVICE PIPING	COLD WATER DISTRIBUTION	HOT WATER DISTRIBUTION
ABS Plastic Pipe, SDR (ASTM D2282)	•	•	
ABS Plastic Pipe, schedule 40 or 80 (ASTM D1527)	•	•	
Brass Pipe (ASTM B43)	•	•	•
Copper Pipe (ASTM B42)	•	•	•
Copper Water Tube, Type K or L (ASTM B88)	•	•	•
Copper Water Tube, Type M (ASTM B88)	•	•	•
CPVC Plastic Pipe, schedule 40 or 80 (ASTM F441)	•	•	
CPVC Plastic Pipe, SDR (ASTM F442)	•	•	
CPVC Plastic Water Distribution Systems (ASTM D2846)	•	•	•
Ductile Iron Pipe, cement-lined (ASTM A377, ANSI/AWWA C151/A21.51)	•		
Fiberglass Pressure Pipe (AWWA C950)	•		
Galvanized Steel Pipe (ASTM A53)	•	•	•
PB Plastic Pipe, SDR (ASTM D3000)	•	•	
PB Plastic Pipe, SIDR (ASTM D2662)	•	•	
PB Plastic Tubing (ASTM D2666)	•	•	
PB Plastic Water Distribution Systems (ASTM D3309)	•	•	•
PB Plastic Pressure Pipe and Tubing (AWWA C902)	•		
PE Plastic Pipe, Schedule 40 (ASTM D2104)	•	•	
PE Plastic Pipe, Schedule 40,80 (ASTM D2447)	•	•	
PE Plastic Pipe, SDR (ASTM D3035)	•	•	
PE Plastic Pipe, SIDR (ASTM D2239)	•	•	
PE Plastic Tube (ASTM D2737)	•	•	
PE Plastic Pressure Pipe and Tubing (AWWA C901)	•		
PVC Plastic Pressure Pipe, AWWA C900	•		
PVC Plastic Pipe, Schedule 40,80,120 (ASTM D1785)	•	•	
PVC Plastic Pipe, SDR (ASTM D2241)	•	•	

(1) Piping for potable water shall be water pressure rated for not less than 160 psig at 73F.
(2) Piping for hot water shall be applied wthin the limits of its listed standard and the manufacturer's recommendations.
(3) Plastic piping materials shall comply with NSF 14.

prevent sewer gas (methane) from reentering the building. The vent systems are designed to prevent the siphoning process that occurs in these traps, as well as toilets, from removing the standing water that acts to prevent the gas from forming. If a response is for a complaint of odors, it might be a good idea to ask if any sinks or drains have not been used in a while; the water in the "p-trap" may have evaporated and, if so, can be solved by running some water into the drain.

HEATING, VENTILATION, AND AIR CONDITIONING (HVAC)

Humans have long sought to master their environment. The desire to be warm in winter and cool in summer has resulted in many inventions over the years. Today's buildings can be heated by a variety of methods. The use of steam is still economical for large commercial structures, but it is not used often in residential structures. Hot water baseboard or **hydronic** heating systems are used instead (see Figure 5-7).

Forced air and heat pumps are much in vogue today. Forced air can be heated by either fuel oil or gas. A heat pump is able to remove moisture from the atmosphere and convert it to heated air. The advantage of both these systems over others is that they both utilize ductwork to convey the heated air, which can also be used when there is a need for conditioned air during warm months.

To cool the environment many people today are revisiting an older approach. The use of ceiling fans or whole-house fans is widespread. The only flaw in using fans is they do not remove the humidity.

Most modern air conditioning systems use a refrigerant gas, which is pumped through coils. Air is made to flow over the coils; moisture is removed and drained away and the chilled air is sent by fan through ducts to the rooms of a building. A return air duct or the space above a dropped acoustical tiled ceiling (the plenum) is used to keep air moving. The constant cycling of air allows for displacement and refill of the conditioned air.

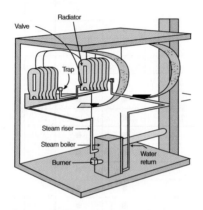

(a) Steam boiler: In a steam system, steam is produced in a steam-rated oil or gas boiler, circulated through insulated pipes to room radiators, and condensed in the radiator, giving up its heat of vaporization. The condensed water then drains back to the boiler for reheating.

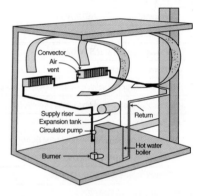

(b) Hydronic boiler: The hydronic, or forced hot water, system heats water in a gas or oil boiler and circulates it through loops of pipe to distribute heat to separate heating zones.

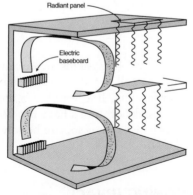

FIGURE 5-7 Typical residential heating systems. (a) Steam boiler; (b) hydronic boiler; (c) electric baseboard and radiant panels; (d) warm-air furnace.

(c) Electric baseboard and radiant panels: Electric-resistance baseboards and radiant ceiling panels convert electricity to heat with 100 percent efficiency; no heat goes up a flue.

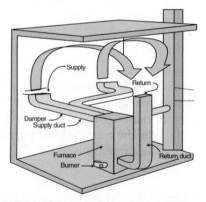

(d) Warm-Air Furnace: In a modern warm-air furnace heat is produced by clean and efficient combustion of gas or oil; the warm air is distributed evenly throughout the building by a blower, supply and return ducts, and registers.

Freon gas (Freon-11 and Freon-12) was used exclusively in refrigerant systems and as a propellant for aerosols until January 1996 when it was banned from all use because it was thought to deplete the ozone layer. Freon is a hydrochloroflourocarbon (HCFC), a subset of chlorofluorocarbon (CFC). CFC is an organic compound that contains carbon, chlorine, and fluorine, a volatile derivative of methane and ethane.

Firefighters may still encounter Freon in appliances and automobiles manufactured prior to January 1996, as well as in refrigerant and air handling systems in commercial and public buildings with older systems. Freon becomes phosgene gas when heated, a major safety hazard for firefighters. Phosgene gas is heavier than air and thus will collect in lower portions of a building. Phosgene gas may appear as a colorless or white to pale yellow cloud. It will smell like newly mown hay or green corn. Do not rely on smell to identify phosgene gas because the gas will attack the olfactory senses quickly. The best course of action is to always wear your SCBA and ensure that air monitoring includes checking for phosgene.

HVAC system problems that result in a call to the fire department can most often be handled by turning off the system and informing the occupant to call a serviceperson. This lowers liability for the department and also brings in a trained professional; many systems are complicated, with microchips, sensitive valves, or programmable thermostats.

INSULATION

Before humans could use any system they needed to stabilize the temperature differential between the outside and the inside of the building. Insulation greatly enhances the properties of HVAC systems and also makes them much more economical to operate. Many materials have been used as insulation over the years. Since the early 1990s these have included cellulose, mineral wool, straw panels, and fiberglass, among others.

Cellulose is the oldest insulating material used in buildings in the United States. Its earliest forms were made of cotton, straw, corncobs, sawdust, and hemp. In the 1970s recycled newspapers became the primary ingredient used. By the mid 1990s, however, cellulose was less in demand by the building industry due to concerns for fire safety. Cellulose is making a strong comeback today. It consists of 75% to 85% recycled paper fiber, usually post-consumer waste newsprint. This makes it a perfect product for the new green building movement as it helps remove paper refuse from landfills. The remaining 15 percent of modern cellulose insulation consists of a fire retardant such as boric acid or ammonium sulphate. The boric acid treatment also helps prevent rodent infestation and inhibits mold growth. The insulation can be installed in walls, ceilings, or floors and can be installed in wet or dry form.

Mineral wool refers to three types of insulation that share common properties. Glass wool (fiberglass) is made from recycled glass; rock wool is made from basalt, an igneous rock; and slag wool is made from steel mill slag. The manufacturing process for the three is similar. The raw material is heated to 1,600 degrees or more and then the liquid is spun to create fibers. These fibers are then intertwined to form batts (panels) or are applied in a liquid form; the material insulates against heat and noise. The main advantage of mineral wool is its ability to withstand fire. It will, however, melt if exposed to high enough temperatures. Fiberglass is the most widely used of the mineral wools for residential applications. Rock and slag wool are used principally in commercial and industrial structures and have replaced asbestos for such applications as pipe and machinery coverings.

Straw panels are used as part of the green building movement. Straw has been used as insulation, as a component in walls, and as a roofing material since colonial times and is now making a comeback in the U.S. market. The main component is wheat straw. A by-product of the wheat production cycle, this straw can be molded into panels or baled and used as a structural supporting wall system. The bale system employs stacking the bales and covering them with stucco, gunite (shotcrete), or cement. Straw is also being partnered with recycled tires, cardboard, and plastics; when used in the manufacturing of panels, the panels can be used structurally as an interior wall covering.

ELECTRICITY AND LIGHTING

There is a danger associated with electricity, but only if it is misused. Electricity travels from the transmission source either by overhead or underground lines. It is transferred to the structure by way of a line called the service drop. If the power is delivered by overhead lines it will culminate at the structure in an assembly consisting of a weather head and drip line, which protects the power line from weather such as rain, snow, or ice. The power continues down a cable referred to as the service entrance to the meter (see Figure 5-8). As it enters the meter, the power contains 240 volts; it energizes the inside service box, which is rated by the amperage available for use. The service then is then stepped down, or split, into 120 volt service, which is common for U.S. residential service needs. If the power is transferred to the structure underground there will not be a drip line or weather head.

There are three primary terms used in discussing electricity: volts, amperes, and watts (volts × amperes = watts). **Volt** is the unit that measures the potential difference in electrical force, or pressure, between two points on a circuit. **Ampere (amp)** is the unit used to measure the amount of current—that is, the number of electrically charged electrons—that flows past a certain spot per second. **Watt** is the unit of power; it indicates the rate at which a device converts electric current to another form of energy, either heat or motion, or the rate at which the device uses electricity. If the circuit is 120 volts and the device requires 4 amperes of service, then the electrical needs equal 480 watts. To find the amperes, divide the watts (if known) by the voltage of the circuit. If a device uses 1,200 watts and the circuit is 120 volts, then the amount of amperes is 10. To mitigate

volt ▪ The unit used in measuring the electrical pressure causing the current to flow.

ampere (amp) ▪ The rate at which a given quantity of electricity flows through a conductor or circuit.

watt ▪ A unit of electrical power (amps × volts = watts).

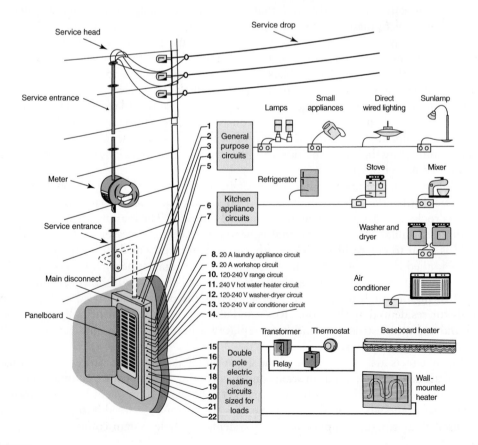

FIGURE 5-8 Electrical service entrance and distribution.

an electrical problem such as a smoking outlet or sparking light, just turn off the circuit to that device. If that poses a problem, turn off the box by pulling the main disconnect handle. If you want to ensure that power has been cut off to a building, have the power company pull the meter head.

Lighting takes many forms within a structure and is simple to use, repair, or cut off. Most appliances are powered by 120-volt 15-amp circuits. Occasionally an appliance may require 220 volts to operate. This should be indicated by a unique plugged outlet in the room; usually the corresponding circuit will require two circuit breakers.

Firefighters should use caution if they suspect that power has been pirated. This involves illegal connections from a pole or another structure and is usually done to circumvent the power company. This may create a totally unsafe connection that exposes firefighters to massive shocks even if they have pulled the meter head.

Commercial

Commercial structures such as office buildings, high-rise or not, require large electrical, HVAC, fire protection, and plumbing systems, as well as dedicated areas to accommodate these systems for distribution throughout the structure. Depending on the building's design, these dedicated areas, commonly called machine rooms or pump rooms, are located one or two levels below grade or street level, or in the penthouse.

PLUMBING

The water requirements for large commercial office buildings or apartments are substantial and necessitate much larger systems than are used in residential construction. By convention, most cities will use differing paint colors to indicate what is inside piping or tanks in machine or pump rooms. The hotter the contents within the piping, the brighter the color it is painted. For example, most steam piping will be painted orange (see Figure 5-9). Heat systems will be yellow, lime green, or even red, though red is usually reserved for fire protection (see Figure 5-10). Water, including that used in air conditioning, is usually color-coded blue or white (see Figure 5-11). Sewage lines are either brown or black (see Figure 5-12). Verify that this is accurate in your own area.

PLUMBING SYSTEM MATERIALS

Copper (type K or L) is used extensively for supply lines; cast iron or ABS is used for waste. Black pipe or plain steel is still used predominantly for sprinklers and standpipes used for fire protection. Copper and plastics are also being used.

FIGURE 5-9 Steam lines in high-rise pump room.
Source: www.shutterstock.com.
Copyright: Kondrashov Mikhail Evgenevich

FIGURE 5-10 Heat systems in high-rise/commercial structure.

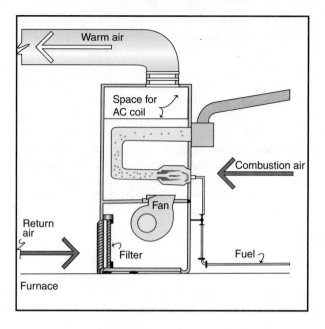

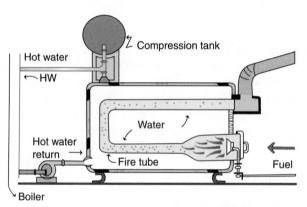

The water meter is usually located in subbasements of high-rise buildings and near a corner of a main road in commercial occupancies (see Figure 5-13). Fresh water is brought into most buildings from city water mains at about 50 psi. This is adequate for most two- or three-story buildings, as garden hoses require 30 psi and are usually on ground floors. Fixtures in public buildings require 20 psi (this includes flush valves). Pressures of 5 to 8 psi are required as a minimum for lavatory faucets and tank-type water closets (toilets). Under static conditions (no flow and zero pressure at the top) 50 psi will sustain 115′ of water [50 × 2.3 (ft of head, the pressure exerted by the weight of water above a given point, per psi)]. This is adequate for an 8- to 12-story building, however, friction loss occurs even in large pipes and at moderate use rates. These practical considerations limit most up-feed systems to around six stories, with four stories considered ideal.

The use of suction tanks is employed to maintain the required quantities of water needed at the pressures required for buildings taller than six stories, or when the water supply from the mains is less than 50 psi. These wooden or steel tanks are usually installed on the top floor or roof and contain as much as 10,000 to 20,000 gallons of water (see Figure 5-14).

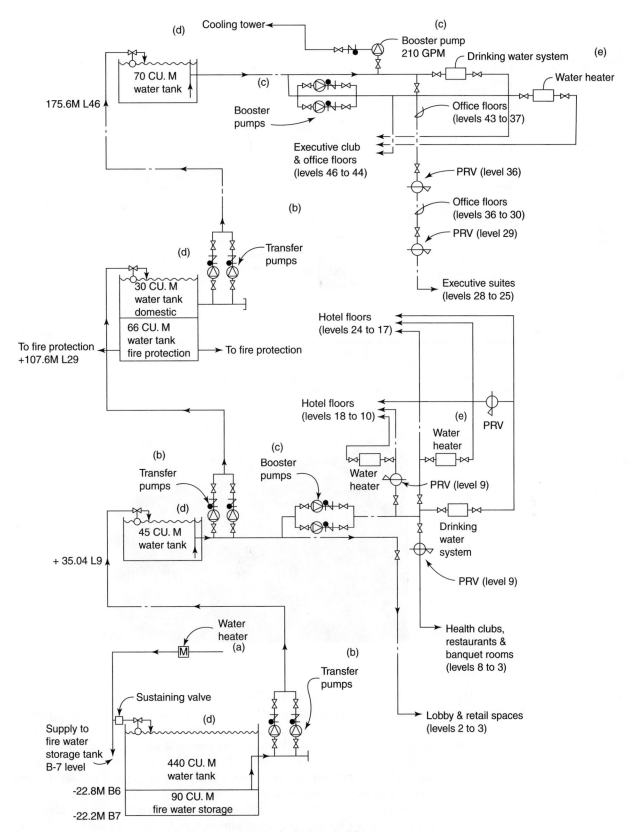

FIGURE 5-11 Example of water lines in machine pump room for high-rise buildings. *Source: Tao, William K. Y.; Janis, Richard R., Mechanical and Electrical Systems in Buildings, 3rd Edition, © 2005, p. 251. Adapted by permission of Pearson Education, Inc., Upper Saddle River, NJ.*

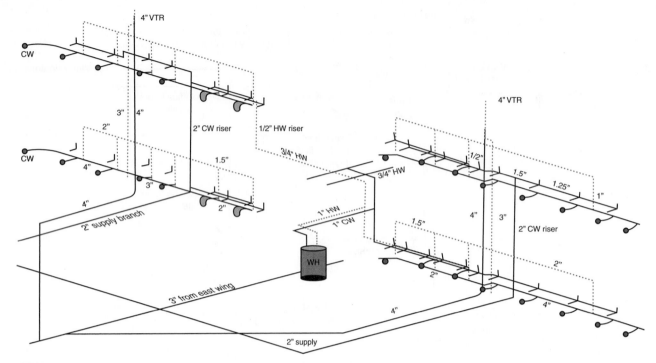

FIGURE 5-12 Sewage lines can be either black or brown in commercial buildings or high-rise structures.

FIGURE 5-13 Typical water meter for commercial occupancies. *Source: www.shutterstock.com. Copyright: Aaron Kohr*

The weight of this water is allowed to flow downward; thus pressures are maintained on the upper floors. Check valves are used to limit the pressures gained from gravity and so maintain the required psi. The lower portions of the tank are dedicated for fire suppression; only the upper portions are available for domestic use. Water is pumped up to the tanks, usually during nonpeak use periods. Float valves are installed to maintain the quantity needed. These tanks require tremendous load-bearing capabilities on the structures. These are commonly called downfeed systems.

The use of pumps to circulate water throughout these structures is mandatory for good performance. The hot water, for example, must travel from the cold side of the system to be processed at a boiler or tankless heater and then pumped up to the highest point of the riser (with the pipe vertical) for use. Unless required by code, as it is in New York

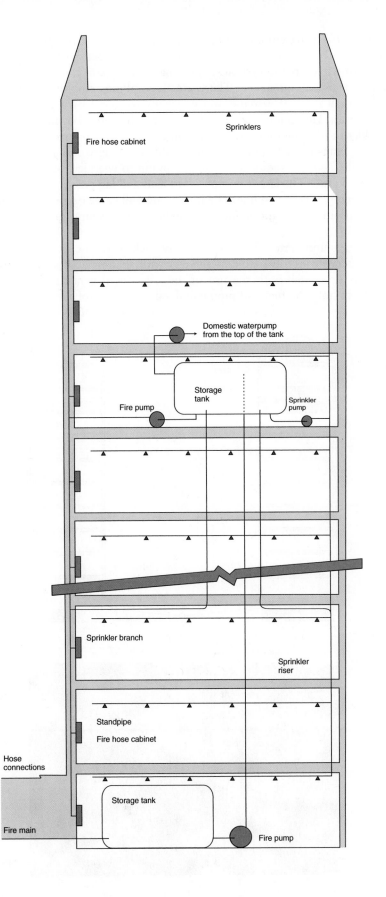

FIGURE 5-14 Typical suction tank.

Sprinklers

Fire hose cabinet

Domestic waterpump from the top of the tank

Storage tank

Fire pump

Sprinkler pump

Sprinkler branch

Sprinkler riser

Standpipe

Fire hose cabinet

Hose connections

Storage tank

Fire main

Fire pump

City, many buildings are employing upfeed pumping systems to replace the suction tank (see Figure 5-15). Because of the weight of the tank it is expensive to install and maintain.

The use of triplex pump systems is proving to be as effective as suction tanks without the weight and maintenance. There are three pumps employed. The jockey pump responds to small rates of demand and is usually triggered by a pressure sensor at the base of the riser. This pump runs until it has reached its maximum delivery rate. An open sprinkler head would initiate the response of a jockey pump (see Figure 5-16), one of the larger pumps to kick in; this pump would continue to run until demand has been met or until the demand requires the second pump to energize, after which they would both continue to run. The larger pumps act as lead and lag positions to prevent the main pump from failing to come on line. The lead pump will come on first, followed by the lag pump. These positions are automatically switched every twenty-four hours automatically (see Figure 5-17).

The triplex pump system has one big drawback; it is susceptible to power failures. There isn't reserve water available as there is with suction tank systems. Buildings can overcome this drawback with the use of diesel-powered generators. These generators require tanks to hold the fuel and piping and valves to control the flow.

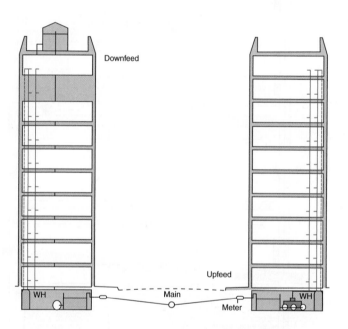

FIGURE 5-15 Downfeed and upfeed systems.

FIGURE 5-16 Typical jockey pump.
Source: www.shutterstock.com.
Copyright: ekipaj

FIGURE 5-17 Main pumps for fire protection.
Source: www.shutterstock.com.
Copyright: ekipaj

Solid and liquid waste present many challenges for builders, engineers, and architects. The system must adequate to the task of moving many tons of effluent from the building to the sewer lines. The weight of the effluent is employed by the system. The waste stacks are larger than in residential systems, but the bulk and weight of the mass act as a plunger to push the effluent down through the system. The horizontal branch lines, often servicing two adjacent bathroom areas, require the most consideration. To maintain the ¼″ per foot fall required and yet remain hidden the piping is installed above or below a floor by dropping the ceiling or installing a raised floor system. If the lowest point serviced by the waste system is below the street level interceptors and ejector pumps must be used. These systems receive the mass and then pump it back up to the sewer lines under the street.

All commercial assemblies require venting to prevent siphoning and to remove residual gases from the system. Venting is configured according to codes and will pass through to the roof.

ELECTRICITY

A battery made up of one or more cells is the simplest source for supplying electricity. A cell usually consists of a carbon rod (positive electrode), and a piece of copper or zinc (negative electrode), and a chemical compound (electrolyte). The electrolyte readily breaks down into positively charged and negatively charged ions, which attach themselves to the electrodes and impart a positive or negative charge. Batteries are either wet cell or dry cell. A wet cell battery uses sulphuric acid as an electrolyte. A dry cell uses an acid paste. Batteries produce small amounts of electricity at low voltages. Generators are required to create larger amounts of power.

Most cities' electricity is generated at a central station. The power is then transmitted either overhead or underground to transformers, which either step down or step up the power. In most residential and some commercial applications the transformers normally step down the power to the 220 volt range. For most large office and apartment building use the voltage will be higher. The power comes into a building at the service drop. A residential structure or small apartment or commercial building usually requires a three-wire drip loop, which is often found on the exterior of a building (see Figure 5-18).

Large commercial structures require a lot of power. Electricity delivered to these buildings is further processed in vaults (see Figure 5-19) because voltages at this level produce great amounts of heat and force. The control lever must be pumped to provide adequate force to shut off power to one of these vaults. Arcing can occur, so this operation

FIGURE 5-18 Service entrance equipment.

1. Service drop
2. Service head
3. Service entrance conduit
4. Threaded hub
5. Meter base (socket)
6. Service entrance panel
7. Main disconnect (main breaker)
8. Grounding bushing
9. Equipment bounding jumper
10. Main bonding jumper
11. Grounding electrode conductor
12. Service entrance conductor
13. Ground/neutral bus
14. Grounding electrode
15. Metallic cold water service
    (Not required if plastic)

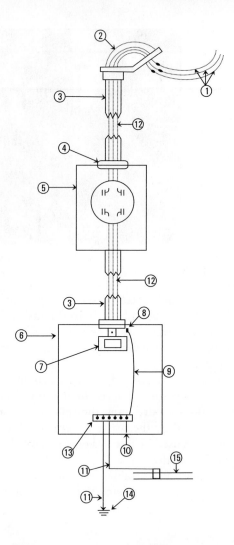

FIGURE 5-19 Vault room in large building.
Source: www.shutterstock.com.
Copyright: emel82

is best left to the building staff and the power company. In most urban cities the vault is below grade at the exterior wall and the sidewalk. It is often covered by grates that further assist to remove heat (see Figure 5-20).

The power is then transmitted to generators that can produce AC or DC power. DC generators are actually AC generators containing a commutator. AC power systems

are replacing many DC systems except for elevators and escalators. DC current flows in only one direction in a circuit. AC current cyclically alternates between poles.

Obviously the large amounts of power and the many potential contact points, switches, motors, wiring, raceways, and busways preclude any use of water in vault areas or motor and pump rooms.

Do not assume that power has been interrupted until a member of the power company has tested the circuits.

Power leaves the transformers and is fed throughout the building. Electric rooms are placed on each floor to handle its service needs. The room contains a service panel or panels; it often contains telephone circuits as well (see Figure 5-21). Remember that although the panel has a shutoff switch, the tower, or the conduit supporting the panel, is energized (hot) because it is being fed from below. Most often the lighting circuits in each room of a building will be higher voltage than the outlets.

Because of the widespread use of computers a power failure can be disastrous, so battery backup is common. This system is referred to as uninterrupted power supply (UPS). Most often the system consists of a Telco (telecommunications) room filled with banks of batteries on metal racks. These rooms can be dangerous for firefighters because they are energized even when you have cut power to a floor. During a fire the batteries may be off-gassing dangerous fumes; cutting the power cuts the ventilation, allowing gases to accumulate. It's best to know where these rooms are (see Figure 5-22).

FIGURE 5-21 Service panel.
Source: www.shutterstock.com.
Copyright: gornjak

BUILDING GREEN

Many residential and some commercial occupancies are adding photovoltaic cells. Such occupancy uses less municipal power; in some cases they will have a surplus, which is sold to the utility company.

These cells generate direct current. A single cell will not generate much power so multiple cells are assembled together to form an array (bank) of cells. The array is called a *solar panel*. The solar panels are assembled together to form an even larger power-producing array. See Figure 5-23 for a typical solar panel installation on a residential building. Typically, for every 100 watts of power generated the system will need approximately 400 square feet of space. The panels can be standalone by using a pillar and assembly in the yard. They can also be installed on the roof of an adjoining structure, such as a patio or garage, and still be attached to the system inside the house. As long as there is unobstructed sunlight the system will produce energy. The panels are connected to the roof of the structure by conventional attachments. See a local contractor for the methods of attachment used for the buildings in your area.

A system can be tied to the municipal power grid. In remote locations they can stand alone through the use of batteries. Units attached to the municipal power grid use an inverter device to convert DC (direct current) to AC (alternating current). See Figure 5-24 for an illustration of how an inverter is used to convert DC to AC current when the system is connected to the municipal power grid. There must be an external disconnect by the meter that allows the system to be disconnected from the grid during power outages (see Figure 5-25). Inside the external disconnect there will be AC and DC breakers that turn off the system during maintenance and emergencies. See Figure 5-26 for an inside assembly showing circuit breakers, panel boxes, inverter, and batteries.

For commercial installations, panel requirements and roof space requirements are great. The weight of the system increases correspondingly. Power needs are much greater for commercial structures. Plan a visit to one of these installations in your area.

Solar panels and photovoltaic systems pose potential hazards to firefighters.

- DC power continues to be generated and the system is live until DC and AC breakers are pulled.
- Roof can be at critical support mass when a system is installed.
- Vertical ventilation cannot be accomplished with solar panels in place on the roof.
- Batteries will produce hydrogen.

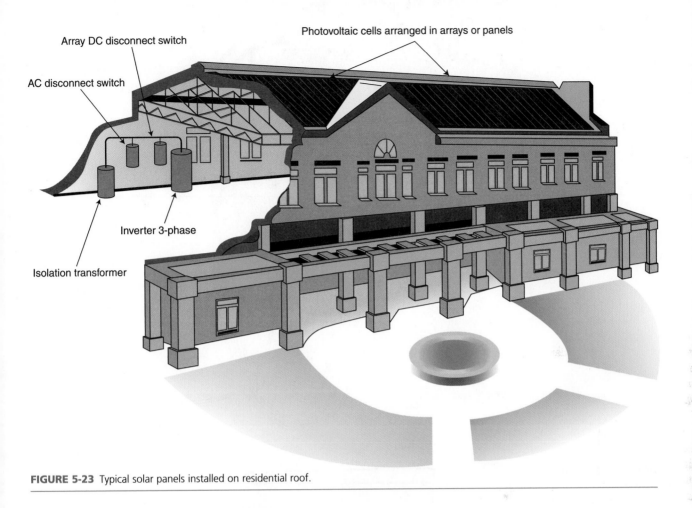

FIGURE 5-23 Typical solar panels installed on residential roof.

HVAC

High-rise buildings need to be economical to operate. Thus the effects of the heat given off by the occupants, computers, lighting, and other sources must be factored into the system in order to effectively handle temperature demands and seasonal changes.

Heat can be provided via circulated water (a hydronic system) or forced air. When the entire building is serviced from a central system, boilers are used to generate hot water and/or forced air for heat; cooling is provided through a central air conditioning system. In commercial buildings with multiple tenants a mechanical room can be located on each floor. The mechanical room houses air-handling units for heat or air conditioning. Central systems are used to collect air from the outside and from within the structure and process it through cooling towers. The cooling towers are usually located on the roof or penthouse (see Figure 5-27). Water flowing over vanes removes the heat from the air being processed. The air then travels to chillers where refrigerant is pumped in and around the piping to further condition the air, which is finally distributed back out through the building (see Figure 5-28).

Duct size and pump size are calculated to give maximum flow with minimum noise. Many of today's systems are controlled by computer-assisted switches and valves.

FIGURE 5-24 Inverter.

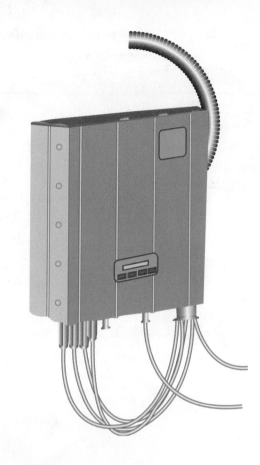

FIGURE 5-25 External disconnect.

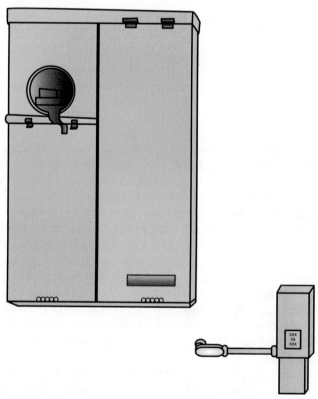

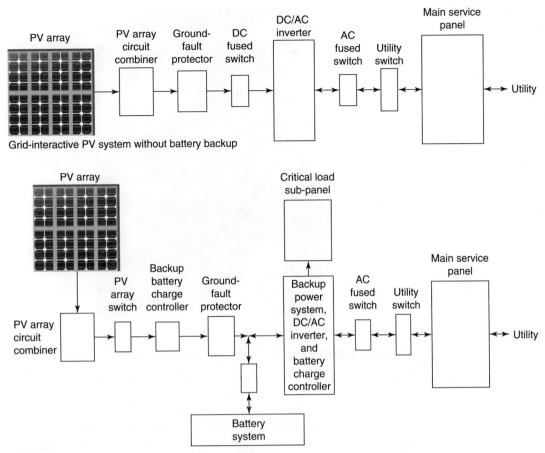

Grid-interactive PV system without battery backup

Off grid system with battery backup and charging assembly

FIGURE 5-26 Interior assembly for voltaic system.

FIGURE 5-27 Cooling tower for high-rise system.
Source: www.shutterstock.com.
Copyright: David Gilder.

FIGURE 5-28 Chillers located in machine room of high-rise. *Source: www.shutterstock.com. Copyright: Damian Palus.*

Smoke Movement in Large Buildings

When a fire breaks out in a building, one of the by-products of this process of combustion is pressure. This pressure, in conjunction with the thermal impact (heating) of the by-products of the fire, causes the by-products to begin migrating horizontally and vertically unless impediments are introduced. A fire also creates gases, including carbon dioxide, carbon monoxide, nitrogen oxides, sulfur oxides, hydrogen cyanide, and hydrocarbons among others. All of these gases are lethal to humans.

The mass of unburned or incompletely burned particles is inherently buoyant. It will rise vertically as long as there is sufficient heat to influence its movement. As it cools, it will begin to stratify and fall. As it is moving it will penetrate any orifice or opening that it encounters. Such openings include stairwells in which a door has been left open, elevator shafts, and plenum areas used as return air ducts. The gases will also penetrate unsealed joints between the floor and exterior curtain walls.

Smoke is influenced by the air currents existing in a building, currents caused by fans or by stack effect. Stack effect, which has a significant influence on the movement of smoke in taller buildings, is the combination of (1) the temperature differential between the inside air of a building and the outside air, (2) the building height (greater than 40′), and (3) the wind velocity on the face of the building combined with the amount of air leakage within the building. For example, if the inside and outside temperatures are the same with little or no wind, the plume will move only in response to the thermal influence of the fire. If the inside temperature is greater than the outside and there is wind, the plume will rise. Conversely, when the outside temperature is greater than the inside temperature the smoke plume will fall even if there is wind.

Firefighters should not automatically assume that a few floors below the reported fire will be clear of smoke. Be prepared to go on air if the smoke is meeting you at the elevator door.

HVAC CONTROL DURING EMERGENCIES

A long-held misconception is that the fire department, when called to an emergency, can regulate all HVAC operations to manage smoke movement. This may be true under some circumstances, but it should not be taken for granted or included in standard operating procedure guides unless confirmed by actual tests. Upon activation of a fire alarm, the HVAC system may be capable of several things.

■ The fire floor can shut down completely with no air movement.
■ The fire floor can shut down while allowing the floors above the fire to exhaust while the floors below the fire intake.
■ Or, the fire floor plus three floors above and two floors below can shut down.

The primary objective for smoke control is to:

■ Maintain a tenable environment allowing the occupants time to evacuate.
■ Control and reduce the migration of smoke between the fire area and adjacent spaces.
■ Provide conditions allowing emergency response personnel to conduct search and rescue operations and to locate and control the fire.
■ Assist with post-fire smoke removal.

There are generally two types of smoke control systems. The dedicated smoke-control system is intended for the purpose of smoke control only. It has separate air moving and distribution systems that do not function under normal building conditions. The non-dedicated smoke control system utilizes the components of the building's main HVAC system. During an emergency the system is designed to change its mode to achieve smoke control.

Whichever system is within the building, there should be a smoke control station provided for firefighter use. This station would be designed to allow for manual override of all systems necessary to counter the situation. Override switches can be found in a variety of places, including in the engineer's office below grade and the fire control console. Once again, test and confirm prior to an actual emergency.

FIRE PROTECTION

Fire protection for large buildings means more than just having a sprinkler system. It includes pumps, piping, tanks, valves, and, in some cases, hoses available for immediate use. Many cities require the use of a suction tank system to ensure that adequate water is available at pressures needed for all floors of a building. These tanks are often 10,000 to 20,000 gallons in capacity and have gated discharge valves on an attached manifold; the tanks are provided to give the fire department additional sources for water on upper floors. For most buildings, external protection begins with standpipe or sprinkler connections. These connections allow the fire department to pump water under pressure into the system to augment the building's own appliances, standpipes, and sprinkler systems. (see Figure 5-29)

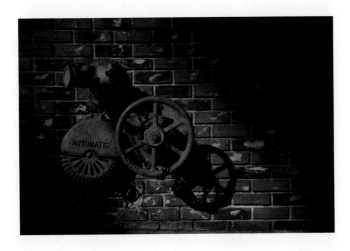

FIGURE 5-29 Fire department connections should be marked. *Source: www.shutterstock.com. Copyright: Worachat Sodsri*

Because the buildings are so large there can be multiple connections, connections that may or may not be interconnected. Careful pre-planning is important to ensure a successful operation when an emergency arrives.

Many buildings use a triplex pump system for providing fire protection. In such a system a jockey pump reacts to minor changes in pressure within the system and either contains the demand or runs until reaching capacity, at which time a larger lead pump energizes. If the lead pump is still insufficient then a lag pump comes online to augment pressures until the necessary level is reached. The lead and lag pumps are interchanged every day to prevent premature wear and tear. These pumps may not actuate in the event of a power failure, a possibility that should be addressed in the pre-fire plan. The valves for fire protection are often gated and secured in the open position. The entire system will be overseen by sensors, switches, and valves that indicate operation or tampering (see Figure 5-30).

Sprinkler systems are designed by fire protection engineers according to applicable codes. Most sprinkler piping is still black pipe although copper, steel, and some plastics are being used in some jurisdictions. Steel is susceptible to rust and corrosion so inhibitors are added if the system is wet. Check valves keep the water in the fire system from entering the domestic system when the systems share supply risers. Parking garages and other unheated areas usually have dry pipe systems. These systems contain air under pressure. If a head actuates, the air is vented and a clapper valve opens to admit water.

The sizes of pipe and the distribution of heads are dictated by code. The type of head and its rating are also prescribed by code. Little support is needed from fire departments except for flow tests and inspection of the heads or paint, dust, or some other contaminant that would prevent actuation. These flow tests and inspections should be scheduled in accordance with your local municipal codes.

Standpipe systems are provided to allow fire departments to connect their hoses to a confirmed water source closer to the fire than their apparatus, which might be many floors below.

There are two types of standpipe systems: wet and dry. Dry systems require the fire department to connect a pumping apparatus to an outside connection. Wet systems are permanently filled and are part of the building's water system; a hose is attached to enable tenants to combat the fire. Firefighters should never use this hose; it can never be relied on for firefighting.

To prevent overpressurization from the suction tank, if present, or from pumps, a pressure-reducing valve (PRV) can be installed between the fire department connection and the riser or between the gate valve and the connection. The key questions for the fire department are if the valve can be set in the field and with what type of tool. An

FIGURE 5-30 Sprinkler valves are often secured in open position. *Source: www.shutterstock.com. Copyright: Brent Wong*

additional question is if it is possible to remove the valve completely if necessary. Learn to recognize these valves and what their limitations might be.

Alarm systems within multiuse high-rises are regulated by local codes. They can be of a variety of types. Gongs, sirens (connected to lights or not), and other systems are available. Many offices and mixed-use buildings have security personnel on duty in the front lobby. The security station will have, among other systems, a fire control monitor. This computer-assisted system monitors alarm activations and in some cases permits security to alert some or all floors. The system may also allow for phone or intercom capabilities between the console and the affected floor. The building may have per-floor phone jacks that are dedicated to fire department use. The fire department must know which elevators work on fire department override and, in case of power failure, whether the elevators return to the lobby or stay in place. The department should also know what run-bys exist for elevators; run-bys are the floors not serviced by an elevator car under normal circumstances.

All capabilities and aspects of a building's systems should be part of an overall incident planning manual. Such a manual is beneficial to incident commanders as well as building staff in case of fire or other system-related emergencies.

ELEVATORS

Elevators are powered in two main ways: traction and hydraulics. The principal apparatus involved in a traction system includes the car, cables, elevator machine, control equipment, counterweights, shaft (or hoistway), rails, penthouse, and pit. Each car is essentially a cage of light metal, with cables attached to the top, supported on a structural frame. Rail shoes fix the car vertically in the shaft. The car is provided with safety doors, operating control equipment, floor level indicators, illumination, emergency exits, ventilation, kick plates, and hand rails.

The cables lift and lower the car. Usually four to eight cables are placed in parallel, the weight of the car distributed equally among them. The cables, fastened to the top of the car by cable sockets, pass over a motor-driven cylindrical sheave at the traction machine (grooved for the cables) and then pass downward to the counterweight, to which they are also fastened with cable sockets.

The elevator machine turns the sheave that raises or lowers the car. It consists of a heavy structural frame on which the sheave and driving motor, the gears, if any, the brakes, and the magnetic safety brake are mounted. The governor, which limits the car to safe speeds, is mounted on or near the elevator machine. The elevator driving motor gets its power from a motor–generator set. This set may be close to the machine or located some distance away, such as the penthouse.

The equipment to control the door operation, starting, stopping, acceleration or deceleration, leveling, and stopping the car consists of the push buttons, contacts, relays, cams, and devices that are operated manually or automatically. In most cases the only manual input is when the occupant directs the car to a designated floor. The balance of the operation of the car is automatic except when the car is in firefighter mode and a key must be inserted within the car enabling firefighters to operate all of the functions of the car.

The counterweights are rectangular blocks of cast iron. The weight is proportional to the weight of the car, resulting in less energy being required to move the car. The counterweights move in the opposite direction of the car; they travel along rails and are held in place by guide rails.

The elevator machine room is usually directly above the hoistway at the uppermost run of the elevator or bank of elevators. It contains the motor–generator set that powers the elevator machine, the control board (which houses the relays and switches controlling the movement and operation), and other control equipment. This room is quite hot during the peak operation of the elevator(s). The control board is likely to be open to allow heat to dissipate (see Figure 5-31).

FIGURE 5-31 Elevator controls in machine room.
Source: www.shutterstock.com.
Copyright: Brian K.

This is a dangerous place for firefighters who are wet or who are not cognizant of the amount of electricity present in the room.

For medium- and high-speed elevators, a DC-powered gearless traction motor is used. The slower-geared traction machines use either AC or DC power at different speeds. Freight elevators most often use the geared machines.

For inexpensive elevator service, hydraulic systems are used. Unlike traction elevators, hydraulic elevators do not need cables, drums, motor–generator sets, or elaborate controllers. The movement of the elevator is controlled by means of a movable rod (plunger) rigidly fixed to the bottom of the car. Instead of a penthouse, hydraulic elevators use a basement machine room. The system requires a much larger motor (there are no counterweights) and a tank for the reclaiming and storage of the fluid used to push the rod. The system acts in much the same way as a car jack: pressurized fluid is forced to the seat of the rod; when the car comes down the fluid is reclaimed and stored in the reservoir (tank). The controls for this system are essentially the same as for traction-type elevators. Hydraulic elevators are primarily used for low-rise applications under six stories.

SCHOOL OF HARD KNOCKS

I worked for a company that developed disaster plans for high-rise buildings in several major cities. These plans included individual floor plans and all the various building systems, including elevator, mechanical, and fire and water pumps. The knowledge that I gained through this work was invaluable during the many incidents that I faced as a fire commander in high-rise structures. I found that many myths about the systems were also debunked by my experience.

The most pertinent thing I learned was that when high-rise structures begin to fail during an incident I could lose power, communications, access and egress, and the ability to pump water for fire attack. When this happens, it is as if all personnel have lost their tether during a space walk. We need to bring our people in the fire service back 100 percent of the time. When orders are given based only on expectations of firefighter courage, the firefighters will die. They need communications, water, and the ability to escape the danger. Disasters do not have to involve fire: If a building is experiencing a major water leak on an upper floor and the lower floors contain equipment that could be damaged by water, if you don't understand or know where the cutoff valves are the disaster will certainly escalate.

In the fall of 1998 units under my command responded to a call to the House Office Building. Ten sprinkler heads contained the fire, but it took approximately two hours to find the sprinkler cutoff valve, which was located six blocks away in another building. This delay led to unnecessary and extensive water damage within the building. Remember: Federal buildings are not required to be inspected by the local fire department; for this reason the department was not allowed to perform familiarization tours.

Apartment Buildings

Apartment buildings are found in every state and in most cities. Apartment buildings can be defined as a building containing several to multitudes of people who either own or rent the space they live in, space that is contained within a single structure. Most apartments are single-story while condominiums are normally two or more stories. Each has its own entrance either to the exterior directly or to a common hallway. The term garden apartment is often confusing as some define it by its configuration and some by unit access to a hallway or balcony. The correct definition is any apartment building or complex, usually low-rise and one to three stories in height, built around landscaped grounds. This landscaping can vary by region.

Apartment buildings, which have been in use for centuries, were originally referred to as boarding houses or lodgings. During the mid nineteenth century, in cities such as New York and San Francisco, railroad apartments (flats) began being built. This type of apartment is very narrow; most are less than 20' wide. The rooms are often laid out in a linear pattern, connected by a hallway. A variation is an apartment that does not contain a hallway, the rooms being connected by pocket doors (doors installed on a rail and which slide in and out of the wall). The term *tenements* has been mistakenly used to indicate apartments built to house people of low socioeconomic status, but the term actually applies to any apartment building in a run-down state or barely code compliant.

Depending on the strength of local code enforcement and requirements, most new apartment buildings contain sprinklers in, at a minimum, the public spaces, halls, storage areas, and foyers. Low-rise buildings (up to three stories in height) may not contain standpipe systems, but mid-rise (three stories to 75') and high-rise (over 75') will usually contain sprinklered public areas and standpipe systems.

The utility systems in mid-rise and high-rise buildings will be similar to similarly sized structures used for offices or hotels. There will be auxiliary backup power, elevators, fire pumps, and large complex systems for plumbing, electrical, and HVAC. The stair system can either be straight or scissor. The windows in apartment buildings are usually operable while windows in offices and hotels are usually not.

The main difference between apartment buildings/condominiums and hotels/offices is that in the former the tenants have control of their own space—meaning there is no daily monitoring by staff of the conditions within each apartment. Even though the occupants of apartment buildings are not transient, many will not know how to escape during a fire or other emergency.

Some considerations for tactics and strategies are listed below:

- Use pre-incident planning to determine the population (the old rule of counting cars in a parking lot will not work in urban areas because many tenants do not have cars).
- Have a unit designated to charge, connect, and pump water to the standpipe and sprinkler systems.
- Know the limitations or assets of the building's systems to support firefighting activities.
- Anticipate a complex operation of search, rescue, and firefighting that will involve many personnel.
- For fires in storage areas without sprinkler systems, anticipate the presence of hazardous materials and use extreme caution.
- Get emergency medical services (EMS), on-scene as soon as possible; anticipate the potential for mass casualties,
- Limit roof operations for fires on the top floor if the building contains trusses or if the roof is a rain roof, the original roof having been covered over with a replacement trussed system for economic or aesthetic reasons.
- Use of the incident command system with recognized terminology.
- First alarm units should have pre-established assignments. All subsequent units should be staged and directed where needed with proper command entities in place.

Elevator systems that do not require a machine room are now available. These are known as MRL (machine room less) elevators. These elevators are designed so that most of their components fit within the shaft. There is also a small cabinet containing a computer. The traction rope is configured for force multiplication via a complex pulley system; the traction motor moves more rope per distance traveled but works half as hard. Check your local codes to ascertain if MRLs are in your building system stock.

ON SCENE

Fire units are dispatched to a residence after a severe rainstorm. The house measures 40′ × 60′ with a full basement. The owner advises you, the first arriving fire official, that the basement is filled with rainwater but no effluent. There is a full bathroom located in the basement as well as a sump pump.

1. What are the problems associated with using the existing sump pump?
2. What other method could you employ?
3. How would you cut off the electricity to the basement?

Summary

It is important to understand building systems whether in residential or commercial structures. Residential building systems are relatively simple in design; commercial occupancies and high-rises are far more complex. Therefore it is essential to plan for emergencies and to be prepared with essential information regarding how the systems in these buildings operate. Those in charge of operations need to be familiar with elevators, HVAC systems, Telco rooms, and plumbing systems.

Fire personnel should ensure that utility representatives are employed to cut off gas, electricity, or water to commercial buildings. All personnel operating near or in elevator machine rooms, electrical vaults, or pump rooms should exercise extreme caution. Electrocution and burn hazards exist; toxic gases are a possibility when ventilation systems shut down.

If firefighters understand the systems employed in large buildings they are better prepared to utilize those systems in times of emergency, however, they must realize that they should never be solely dependent on the systems. To ensure firefighter safety they must plan strategies and tactics that include a plan for safe withdrawal in case the systems start to fail. Think of high-rises as mountain climbing expeditions: if you didn't bring it then you won't have it. This is true whether the high-rise is vertical or horizontal.

Review Questions

1. Describe some similarities and differences of systems in residential and commercial structures.
2. Discuss the importance of determining the location of bathrooms and kitchens as it relates to firefighter safety.
3. Discuss the relationship of the pumps provided for fire protection versus pumps used for domestic water supply.
4. What are some of the dangers in attempting to cut off utilities in a commercial building?
5. Describe the color-coding system used in pump rooms or machine rooms in commercial structures.
6. What are some of the dangers of using elevators?
7. Discuss ventilating large commercial structures.
8. Discuss alternative ways to supply water if domestic water sources are compromised.
9. Discuss how green building systems influence firefighter operations during a fire.

Suggested Reading

Avillo, A. 2002. *Fireground Strategies*. Saddlebrook, NJ: Penn Well.

Mahoney, E. 2004. *Fire Department Hydraulics*. Upper Saddle River, NJ: Pearson Education.

Clark County Fire Department, Las Vegas, Nevada. *MGM Grand Hotel Fire Investigation Report*.

NFPA 13. 2002. *Automatic Sprinkler Systems Handbook*. Quincy, MA: National Fire Protection Association.

NFPA 92A. 2009. *Standard for Smoke Control Systems Utilizing Smoke Barriers and Pressure Differences*. Quincy, MA: National Fire Protection Association.

NFPA 25. 1998. *Water Based Fire Protection Systems*. Quincy, MA: National Fire Protection Association.

NFPA 5000. 2003. *Building Construction and Safety Code*. Quincy, MA: National Fire Protection Association.

Smith, J. 2002.*Strategic and Tactical Considerations on the Fireground*. Upper Saddle River, NJ: Pearson Education.

CHAPTER 6

Building Construction Types

KEY TERMS

cockloft, *p. 145*

exoskeleton, *p. 136*

internal skeleton, *p. 136*

parapet walls, *p. 145*

parged, *p. 136*

plenum, *p. 135*

OBJECTIVES

After reading this chapter, you should be able to:

- Identify building construction types.
- Define the terms associated with building types.
- Perform more complete size-up reports regarding building types.

Resource Central

For additional review and practice tests, visit **www.bradybooks.com** and click on Resource Central to access book-specific resources for this text! To access Resource Central, follow directions on the Student Access Card provided with this text. If there is no card, go to **www.bradybooks.com** and follow the Resource Central link to Buy Access from there.

Introduction

The evolution of building codes and their enforcement can almost always be traced to disasters. This is especially true with regard to fire. Fire continues to be the biggest killer for occupants of a structure. Chapter 2 looked at the political process of code development. This chapter examines building construction types. The development of Type III construction evolved after the disasters in San Francisco, Chicago, and Pestigo, Wisconsin. The fires had different causes, but there was one common thread that lead to the mass devastation that ensued: the structures were predominantly constructed of wood and were either in close proximity or connected in rows. The disasters led to an outcry from the insurance industry and civic leaders demanding code changes for row-type housing that would prevent the spread of fire from unit to unit.

The significant fires at the Interstate Bank in Los Angeles and Meridian Place in Philadelphia led not only to changes in codes, but had ramifications in terms of command. A building's construction is defined by certain characteristics covered in many building codes by conditions as set forth in the National Fire Protection Standard 5000 *Building Construction and Safety Code*. When responding firefighting units approach a structure they will often list the building type as part of their brief initial report, using nationally accepted fire service acronyms such as WALLACE WAS HOT (C stands for construction type) and COAL WAS WEALTH (once again, C stands for construction type). This will be addressed in Chapter 10.

The problem with this traditional approach is that the construction industry has the ability to mimic or replicate almost any type of construction aesthetic desired; the appearance of a building may not be a true representation of its construction type. For example, one neighborhood in Washington, D.C. consisted of rows of Type III residential homes and businesses; another consisted of ten four-story multi-tenant apartment buildings also constructed using Type III methods. Then the philosophy for housing the poor changed and the apartment buildings were replaced with row homes. The fronts of these newly constructed row homes mimicked the others in the area, with parapets and elaborate facades of wonderfully carved granite, masonry, and stone. Arches adorned the area above windows and doors in the front of the buildings. The sad reality for firefighters was that all these new buildings were not Type III, but Type V with veneers that made them far more combustible and susceptible to collapse earlier.

Construction of this type is not only occurring on the east coast; it can be found in all parts of the country as builders and developers mimic various styles to create whatever atmosphere or look is desired by the potential occupants of their buildings or homes.

It has become increasingly difficult to discern building types solely from the outside of a building. By the end of this chapter, you should be better able to distinguish the type from cues available outside and inside.

Type I

Type I construction is commonly called *fire resistive* and is the most effective for preventing the spread of fire. None of a Type I building's main structural components will support combustion. Fire-resistive buildings can be low-rise, mid-rise, or high-rise. All major public areas will need to satisfy a code requirement for fire rating.

FIRE RATINGS

The term *fire rating* defines the amount of time, in hourly increments, that a component or assembly can withstand heat or fire before failing. A code may require a specific type

FIREHOUSE DISCUSSION

The Station nightclub fire occurred on February 20, 2003. The fire would cost 100 lives; another 230 people were burned, suffered from smoke inhalation, or received crush injuries from being trampled. This fire highlights the importance of fire inspections and code enforcement. It also highlights the complex issues faced by first responders at scenes involving mass casualties and dynamic rescue operations. The Station fire was the fourth deadliest in the United States. The worst was the Cocoanut Grove fire in Boston, Massachusetts in 1942, which claimed 492 lives.

The Station nightclub was built in 1946. It was originally called Casey's Inn. It was constructed of wood (Type V), with sawn lumber as major structural members. The club was 4,484 square feet. The building measured approximately 65' × 75'. The roofing materials were wood shingles over plywood on the front of the building's façade. This façade was attached in shear to the building and ran the length of the front side. The remainder of the roof area was flat with a built-up roof of mopped tar and paper. The exterior walls were framed sawn wood structural members with plywood and board sheathing. The building had burned in 1972 and was extensively damaged, thus the mixture of sheathing materials. These were covered by wooden shingles, which were painted. The interior of the club was covered with wood photocopied paneling over ½" Sheetrock. The insulation on the exterior walls consisted of polyurethane foam inside of the stud chases. Around the stage area the owners had attached a non-fire retardant foam material to the paneling in an effort to reduce noise. The carpeting throughout the building was a polyester blend and non-fire retardant. There was a fire and smoke alarm system installed. There were no sprinklers, although the building was required to have a sprinkler system at the time of the fire. The club had originally been classified as a restaurant, which did not require sprinklers, but then changed occupancy to a nightclub, which did. This fact went unnoticed during the fire inspection a few weeks before the fire.

The club was overcrowded on the night of the fire. It was rated to hold approximately 420 patrons, but the estimated crowd at the time of the fire was over 450 patrons. The club had four egress points. A main door at the front of the building, two side doors on the left side of the building, and a door on the right side of the building adjacent to the stage area. Most of the crowd attempted to exit via the door through they which had entered, which was the front door. Because of the overcrowding and limited egress points, the club could be considered a "target hazard." The term *target hazard*, as applied by the National Fire Academy, designates any building, occupancy, or venue that would overwhelm the responding fire department almost immediately.

The fire department operated four engine companies with an officer and firefighter; one tower ladder truck and a HazMat unit, cross-manned by an officer and firefighter; two rescue units (ambulances) with two EMTs; and a battalion fire chief. The department operated within a statewide mutual aid agreement, which eventually brought over 100 firefighters and officers to the scene. At the time of the fire there was no common operating communications network that allowed all units to talk directly to each other and to a centralized communications center. The first unit arrived on the scene quickly and deployed a 1¾" attack line. A short time later a mass casualty plan was implemented; a mutual aid fire chief took command of the plan. Another fire chief took control of the transportation component. All of the aspects of triage, treatment, and transport were handled expeditiously and competently.

A major portion of the main roof collapsed within forty-five minutes of the onset of the fire. Another portion to the right of the main doors collapsed one hour into the fire. Aggressive interior firefighting was not possible due to untenable conditions, the sheer volume of patient care, and rescue efforts. Approximately 100 people were saved directly by fire department personnel.

The following contributed directly to the large loss of life:

- The installation of flammable foam
- Lack of an installed sprinkler system
- Poorly designed egress paths
- Insufficiently trained staff
- Lack of portable extinguishers

of component, assembly method, or specification code, or it may accept a comparable unit or assembly that can be shown to accomplish the same performance code result. Most building components have an Underwriters Laboratory label attached, signifying that they have been tested and rated for a particular time period.

THINK ABOUT IT!

Most of us attend events at public venues without considering what our escape plan should be in the event one becomes necessary. Fear and self-preservation are two of the strongest primal urges, urges that supersede any code of morality or decency. There are no heroes, only survivors, and the notion of women and children first is quickly cast aside during a crisis.

Even as firefighters, it is incomprehensible to arrive at a scene with the amount of carnage that occurred at the Station fire.

- Using the definition of target hazard, determine if you have any of those situations in your area (either where you live or where you respond).
- Look for blocked exits, chained exit doors, or other situations that would impede your safe egress from a restaurant, club, or other public place. Should you report any deficiencies to the fire inspector?
- Find out the code requirements for sprinkler installation in bars and restaurants in your area. See if there are occupancies that are noncompliant.
- Know where all sprinkler connections are as well as the size of the nearest water main that services those bars and restaurants.
- Make pre-planning for these occupancies a major priority.

Most codes require a minimum four hour fire-resistance rating for all stairwells, bearing walls, elevator shafts, and columns. This can be accomplished through the use of cement masonry units, cinder block, or Sheetrock over rated steel studs. Each layer of fire-rated 5/8″ X Sheetrock carries a forty-five minute rating. An 8″ cement block carries a four hour rating. Elevator shafts are usually constructed of block or poured-in-place concrete. The exterior walls, floors, and roof must be made of noncombustible materials. These will be either poured-in-place concrete or concrete poured over corrugated metal decking, referred to as Q decking. The roof can have a 1–2 fire rating. All doors that access public space must have fire ratings and self-closing devices. Drop ceilings must also be fire rated. If you are performing an inspection on this type of building, the easiest way to verify whether ceiling tile or doors are fire rated is to look for the required seal or stamp. All rated assembly parts must be labeled; the label cannot be painted over, obscured, or reused on another product.

BUILDING GREEN

Relatively new products getting code acceptance are OSB protected panels and enhanced types of drywall. Oriented Strand Board (OSB) is made from recycled wood chips bound with paraffin and isocyanate glue. The panel is coated with a mixture of fiberglass reinforced non-combustible magnesium oxide and crystallized water. The panel is used as a substitute for 5/8″ fire code Sheetrock.

Fire Code C is another gypsum product. The face has a distinctive blue coloration. It is being touted as better than typical X fire code drywall. Yet another product has a rear foil face attached with adhesives and is designed to be installed as a protective layer against moisture and mold.

Ensure that products such as these are approved for use by your local authority having jurisdiction (AHJ). The AHJ is usually your fire marshal or building official.

HVAC SYSTEM AND SMOKE MOVEMENT

The HVAC system may contain smoke dampers and fire stops. The smoke dampers, when installed, activate upon actuation of the fire alarm or smoke management system. The fire stops, as installed, are controlled by fusible links or other devices. The **plenum** area above the acoustical panel or Sheetrock ceiling often serves as the return for the

plenum ■ The space above the line of suspended ceilings sometimes used as the return area for HVAC.

HVAC. This can be a very vulnerable area for fire spread (see Figure 6–1). Sprinklers are actuated from below and discharge below. A fire can communicate between overhead spaces quickly and overpower the system.

EXTERIOR WALLS

internal skeleton ■ Architectural term applied to buildings supported by box-like structures. The exterior walls are curtain walls. Most high-rise build-ings are constructed this way.

exoskeleton ■ A term applied to buildings in which the exterior walls, including sheath-ing, are load bearing.

The exterior walls of Type I structures are constructed of some form of masonry or concrete. Reliable visual clues for this construction are horizontal lines of concrete visible at floor levels and vertical lines visible at regular intervals. These structures are referred to as being either internally or exoskeletally framed. If the building has an **internal skeleton** the support comes from columns and girders (if it is steel) or poured-in-place concrete columns and floor assemblies. If the building has an **exoskeleton** the exterior skin and framing support the structure. The Willis (formerly Sears) Tower in Chicago is an example of an exoskeletal building, as were the World Trade Center towers. The application of curtain walls is prevalent with internally-framed assemblies. As the name implies, curtain walls are not structural, however, they do form a skin that keeps the elements out and the building's controlled atmosphere in (see Figure 6–2).

The builder can use panels or masonry to enclose the structure. The panels can be of any finish that is desired, from granite to marble, or any other noncombustible finish (see Figures 6–3 and 6–4). The panels are manufactured off site and transported to the building where they are welded, or attached in some other fashion, to points placed into the columns of floor assemblies; these points are most often metal plates.

There are gaps between the panel and the floor edge on each floor (see Figure 6–5). The gaps should be sealed with a fireproof material. If masonry units are used, this is usually accomplished with one wythe of clay brick and one wythe of 4″ cement block or one 8″ block **parged** over with grout to form stucco. Parging is the application of a thin coat of cement plaster used to coat a masonry wall to make it more aesthetically pleasing or to assist in making it water tight (see Figure 6–6). Other points for smoke and fire spread are around pipes, conduits, and other utilities that pass from floor to floor; these should be sealed with materials compliant with ASTM E-814, *Fire Tests of Through Penetration Fire Stop*. Fluid seals are also required in hospital operating rooms, schools, laboratories, and universities. The glazing is contained within metal frames. The skeleton of the building is either poured-in-place concrete with rebar and

parged ■ Coating of a masonry unit with a thin layer of plaster.

FIGURE 6–2 Internal skeleton building under construction.
Source: www.shutterstock.com.
Copyright: ekipaj.

FIGURE 6–3 Non-load-bearing lightweight concrete block curtain wall.
Source: www.shutterstock.com.
Copyright: svic.

concrete floor decks or steel. Steel can be used if it has been sprayed with fireproofing (see Figure 6–7). In the last twenty years, the use of cementatious coatings and intumescent paint has replaced earlier forms of asbestos, mineral, or paper fire coatings for sprayed-on fire proofing.

FIGURE 6–4 Vertical
and horizontal control
joint between precast
brick curtain walls
*Source: Andres, Cameron K.;
Smith, Ronald C., Principles
and Practices of Commercial
Construction, 7th Edition,
© 2004, p. 468. Reprinted
by permission of Pearson
Education, Inc., Upper Saddle
River, NJ.*

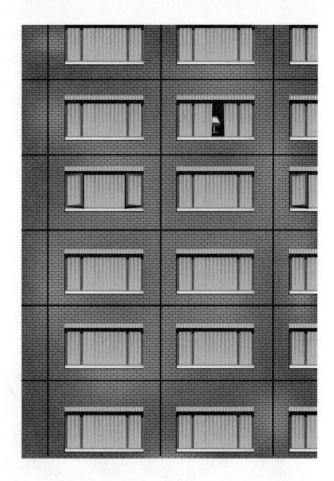

ROOF ASSEMBLY

The roof can be either concrete or concrete over Q decking, provided the decking has been fireproofed. The roof covering can be built up or have a rubber membrane. The latter type of roof covering is gaining popularity as it lasts longer than built up and requires far less maintenance (see Figure 6–8). Expect to encounter many more buildings incorporating green roofs.

Although ultimately dependent on the building code, sprinklers are generally present throughout Type I constructed buildings, especially if they are used for office or apartment space. Many cities still do not require older buildings to have sprinkler systems. For many urban areas with high-rise buildings, there is political motivation to allow current buildings to wait on sprinkler compliance. A code is likely to specify that when a certain percentage of an existing floor is renovated the entire floor must then have sprinklers installed. It always seems that the renovation does not meet or exceed the limit. The foolishness in this situation is that a renovation is the ideal time to retrofit and add a building system, such as sprinklers. The price will be more economical and the system can be better designed.

The use of Sheetrock throughout a building adds to its fire resistance. When the millwork (trim, wood framing members for the installation of cabinets, and so on or other wood assemblies not considered part of the furnishings) is installed it must be treated with either a fire-resistive coating or, in the case of framing members, injected with a salt that gives it a fire-resistive characteristic. The rule of thumb for Sheetrock is forty-five minutes rating for every $5/8''$ layer, fifteen minutes if $1/2''$ is used. Therefore three layers of $5/8''$ Sheetrock yield a two hour rating. Sheetrock installed on lightweight steel studs provides a reliable barrier and is easy to breach if tactics call for a flanking attack. A flanking

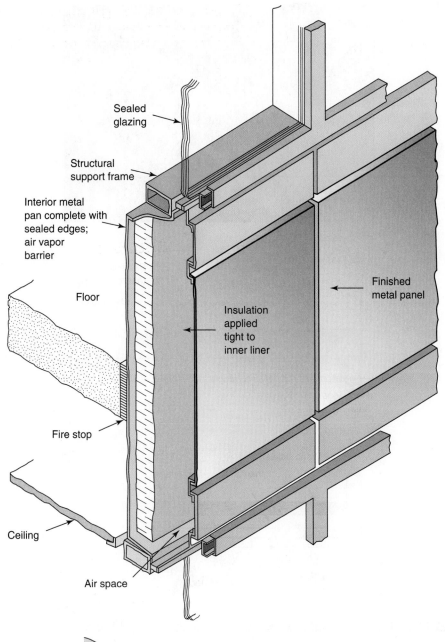

FIGURE 6–5 Lightweight sandwich curtain wall. *Source: Andres, Cameron K.; Smith, Ronald C., Principles and Practices of Commercial Construction, 7th Edition, © 2004, p. 474. Reprinted by permission of Pearson Education, Inc., Upper Saddle River, NJ.*

Sealed glazing

Structural support frame

Interior metal pan complete with sealed edges; air vapor barrier

Floor

Fire stop

Ceiling

Air space

Insulation applied tight to inner liner

Finished metal panel

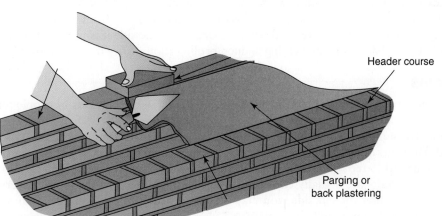

FIGURE 6–6 Applying parging over masonry.

Header course

Parging or back plastering

FIGURE 6–7 Structural steel beams and decking fire protected with sprayed-on insulation. *Source: www.shutterstock.com. Copyright: Stephen B. Goodwin.*

FIGURE 6–8 Protected membrane roof on concrete and steel roof decking. *Source: Andres, Cameron K.; Smith, Ronald C., Principles and Practices of Commercial Construction, 7th Edition, © 2004, p. 426. Reprinted by permission of Pearson Education, Inc., Upper Saddle River, NJ.*

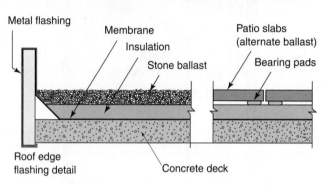

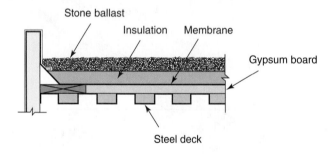

attack is conducted by coming at the fire from an angle instead of head on, which is the traditional aggressive interior mode. This can be accomplished from both sides of the fire or, if necessary, from only one side. If attacking from one side it is preferable to have the wind at your back, thus preventing sudden proliferation of fire if the windows fail.

STAIRS

Buildings constructed for public assembly or to contain large numbers of people for commercial or mercantile reasons conform to NFPA 101 *Life Safety Code* for areas of public access. The stairs can be of a variety of configurations; scissor and U-returns are commonly used. The stairs can either be poured in place or fabricated off site. Off-site fabricated stairs are connected to the floors either by welding or bolting (see Figures 6–9 and 6–10).

Scissor stairs are actually two sets of stairs in the same shaft that service every other floor. If you enter the left side, you may only have access to odd-numbered floors; from the right side you may only have access to even-numbered floors. This can be a problem

FIGURE 6–9 Attaching stair assembly by welding.
Source: www.shutterstock.com. Copyright: Olga Dmitrieva.

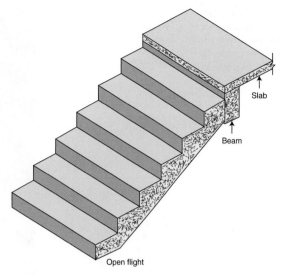

FIGURE 6–10 Concrete stairs.

for firefighters. The standpipe system will often alternate between stairs on alternating even-odd floors. Pre-plans for this situation are a must.

U-return stairs have the landing at a mid point between floors; stairs will service the floor above and below with the landing as the transition point.

Type I buildings pose significant issues for firefighters because of their height, size, and population density. The height of a high-rise building poses problems for firefighters because they have to carry all of the equipment necessary to perform operations into the building. Think of high-rise fires as mountain expeditions: if you don't have it cached or staged ahead of time then you must carry it in. This process requires significant numbers of personnel and will certainly lead to delays in performing the firefighting attack. If the building loses power you are going to have a significant problem maintaining support for an aggressive interior attack because you won't have the use of elevators, the fire pumps may not run, lighting will be limited to battery-powered exit lights, and you will need to rely solely on the fire department's communications system, which may be ineffective.

In low-rise and mid-rise Type I structures the magnitude of the floor plan does not complement your efforts. The amount of time elapsed while personnel are in the IDLH (immediately dangerous to life or health), environment will control how successful operations will be. There are no average sizes for these buildings, but it is safe to state that to search, rescue, and perform fire attack operations will require significant supplies of air that will need to be constantly replenished. Firefighters will become lost or trapped if not successfully roped off.

The deployment of manned or unmanned large-caliber devices in these buildings will make the operation safer.

Type II

Type II construction is similar to Type I in their use of Sheetrock and fire-rated doors. Type II structures are often used as warehouses, commercial facilities, and industrial sites. The ceilings are 20′ or higher from the finished floor. The building can be sprinklered. The assemblies shown in Figure 6–11 are often used when economy is required.

Originally, cast iron trusses were used for the roof assembly, which was often of bowstring design. Most often the earliest forms of this type of construction had exterior siding of masonry or stone or a combination of both.

FIGURE 6–11 Cold-formed girts and roof purlins in a Type II building under construction.
Source: www.shutterstock.com.
Copyright: Richard Thornton.

Main Structural Members

The main structural members must be of a noncombustible material: masonry, steel, or a combination of the two. The masonry units can be concrete, clay brick, or cement block (see Figures 6–12, 6–13, and 6–14). When constructing low- or mid-rise structures, the use of lightweight steel shapes, open web trusses, and lightweight bar joists is prevalent (see Figures 6–15 and 6–16).

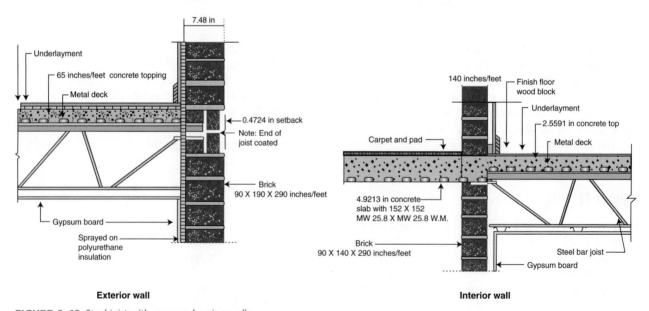

Exterior wall

7.48 in

Underlayment

65 inches/feet concrete topping

Metal deck

0.4724 in setback

Note: End of joist coated

Brick
90 X 190 X 290 inches/feet

Gypsum board

Sprayed on polyurethane insulation

Interior wall

140 inches/feet

Finish floor wood block

Underlayment

2.5591 in concrete top

Metal deck

Carpet and pad

4.9213 in concrete slab with 152 X 152 MW 25.8 X MW 25.8 W.M.

Brick
90 X 140 X 290 inches/feet

Steel bar joist

Gypsum board

FIGURE 6–12 Steel joist with masonry bearing walls.

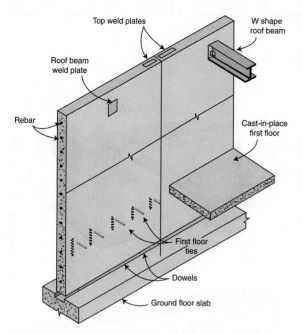

FIGURE 6–13 Load-bearing tilt-up panel with weld plates for roof beams.

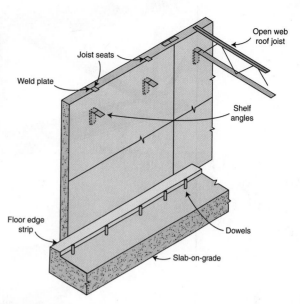

FIGURE 6–14 Load-bearing tilt-up panel with weld plates and shelf angles.

The contents of these structures will often contain large fire loads (the amount of combustible material per square foot of structure per floor). They will also have high rack storage. Firefighters should not be misled into thinking that because the building components won't burn that it will be safe to conduct an aggressive interior attack with handlines.

ROOFS AND FLOORS

The roof and floor members can be lightweight bar joists or steel shapes. These metal trusses are quite effective for carrying the loads imposed on them, but they will fail when

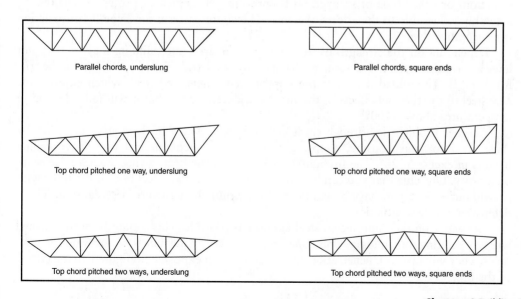

Parallel chords, underslung

Parallel chords, square ends

Top chord pitched one way, underslung

Top chord pitched one way, square ends

Top chord pitched two ways, underslung

Top chord pitched two ways, square ends

FIGURE 6–15 Steel joist types.

FIGURE 6–16 Types of joist ends.

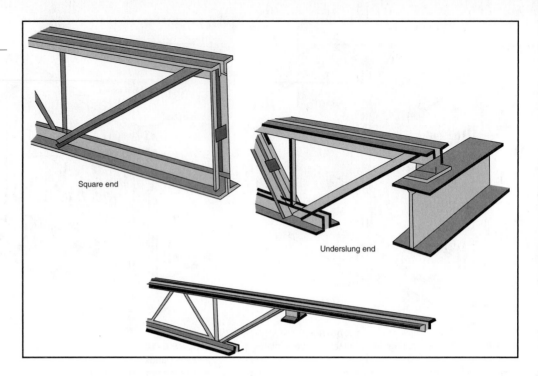

Square end

Underslung end

exposed to temperatures of approximately 1000°F. The connections can be welded or riveted or can use shear-resistant bolts with nuts.

The floor assemblies can be concrete over metal Q decking. The roof will be metal decking or fireproofed plywood. The injection of salts is used to treat the plywood. When the use of combustible plywood or OSB is allowed by code, there has been an initiative to install sheets of 5/8″ Type X Sheetrock over the roof support and then install the roof sheathing. This assembly method is proving to be successful at preventing fire spread in multiple occupancy structures such as apartment buildings and garden apartment complexes. The exterior can be masonry, steel, concrete, or stone. The steel will not be protected; it will not have a gunite (a material used in swimming pool construction) or other type of sprayed-on fireproofing. This lack of protection makes the assembly vulnerable to the effects of flame. The common term for Type II construction is noncombustible.

Depending on the strength of the code in your area, the building may or may not have a sprinkler system. The same problems associated with Type I buildings will be true for Type II. The added danger is the unprotected, unencased steel, which exposes the raw steel to the thermal effects of the fire. Steel doesn't burn, but it will fail if heated to temperatures above 1100°F.

Another problem for firefighters is ceiling height. Many of these structures, mainly those used for warehousing, shops, garages, or other commercial uses, have ceiling heights in excess of 20″. The firefighters will not feel the heat, giving them a false sense of security. One clue that you can bet your life on is that if you feel heat at the floor with or without smoke you have a significant fire within the structure. Use caution. Think defensively in these situations.

Get water on the unprotected steel as soon as possible. This should be accomplished without entering the building. Unmanned monitors are an ideal method for water deployment. They will give you penetration and reach without exposing firefighters to the effects of the fire.

Type III

Type III is referred to as *ordinary construction*. It evolved as a panacea against conflagrations in urban areas. Many nineteenth- and early twentieth-century buildings are Type III. Type III can be found in a row configuration or in stand-alone buildings. It is best identified by its assembly method, which consists of vertical masonry walls with horizontal wooden floor members embedded in the masonry. The joists are 12″ to 16″ on center and are fully-dimensioned lumber. The vertical walls are usually 3 wythes thick. The end units of a row will have thicker walls to support the lateral load. In a row configuration, the walls adjoining side by side buildings are called party walls. The combined thickness for these walls is 6 wythes. The row can share a common roof with a **cockloft**, the area between the ceiling and underside of roof assembly. Most of these structures are 12″ to 20″ wide and 30″ to 60″ long. Rows can be up to six stories tall. The alpha (front) side will, as a rule, include decoration such as elaborate arches of brick, stone, granite, or marble and may include stained-glass transoms (see Figure 6–17). There will be **parapet walls** along the front roof line (see Figures 6–18 and 6–19) and ornate cornices of pressed tin, wood, or masonry (see Figure 6–20). The parapet walls will be capped with copper, terra cotta tile, stone, or masonry. This is referred to as coping. Parapet walls are very susceptible to full failure, particularly if the fire is in the roof area or on the upper floor in the front of the building.

cockloft ■ The space below the underside of the roof and the top of the ceiling. It is associated with flat roof configurations.

parapet walls ■ The portions of any exterior wall, party wall, or fire wall that extend above the roof line.

FIGURE 6–17 Elaborate fronts with parapet walls are classic Type III cues.

FIGURE 6–18 Parapet wall with coping.

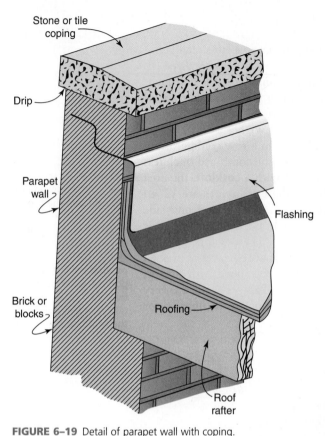

Stone or tile coping

Drip

Parapet wall

Brick or blocks

Flashing

Roofing

Roof rafter

FIGURE 6–19 Detail of parapet wall with coping.

Turrets or towers are often used at the ends of rows for aesthetic purposes and to increase floor space. A tower is a round structure built from the ground up. A turret is a round structure that originates on an upper floor (see Figures 6–21 and 6–22.) Many Type III buildings have cantilevered bump-outs that allow a room to protrude past the vertical plane of the front wall. The vertical studs of bump-outs and turrets are built on top of the cantilevered floor joists (see Figure 6–23).

The roof framing runs from front to back; the interior bearing walls support the roof rafters. The roofing material can be rolled felt with cold tar built up with aggregate (see Figure 6–24), rubber membrane, or, in some cases, standing seam metal. Type III buildings usually have flat roofs, but some are constructed using a mansard style, sometimes referred to as Federal or French style. The use of mansard roofs increases the usability of the top floor for the inhabitants (see Figure 6–25). If the top floor is finished, the roof also serves as the walls. There are often windows or decorative octagonal or circular portals in these vertical walls. The exterior is usually clad with slate tiles.

Many buildings using mansard roofs are found in urban areas such as New York City, Philadelphia, San Francisco, Austin, Texas and Washington, D.C. The choice of roof style can be traced to economics. With its vertical sides, the mansard roof allows for increased space without increasing taxable space. You may find mansard roofs used to hide mechanical systems on the roof. The exterior roof walls are constructed with double-planed pitches, but the interior side is vertical, forming a straight wall surface. The remaining space is open to the environment and was previously used as an early form of intensive green roof; now it is used, for example, to house the exterior HVAC air-handling units.

FIGURE 6–20 Elaborate cornice work can often be found in this type of construction.

FIGURE 6–21 Turrets are constructed on upper floors.

FIGURE 6–22 Towers are constructed from the ground up.

Often skylights are located above stairwells and interior bathrooms in Type III construction. Many of these structures have roof hatches that provide access to the cockloft. The stairs are stacked. Some buildings originally owned by wealthy individuals may have a second set of stairs running from the kitchen to the second floor. The second set of stairs allowed the domestic help to perform their duties without disturbing the owners. They are often sandwiched between a party wall and a bearing wall and are quite narrow and steep.

The earliest forms of Type III construction the floor joist ends were embedded in the vertical masonry walls. An angled cut was applied to the ends to prevent collapse should the floor joist fail. The cut of 30 to 40 degrees, referred to as a *fire cut*, was angled back towards the interior to allow the joist to fall into the structure. Modern applications of this assembly use steel hangers embedded into the masonry; the hangers cup the joist ends. The use of trusses for floor joists has been common for the past thirty years. The party walls are parged with plaster directly onto the brick. The remainder of the interior is plaster over wood lathe or wire mesh; it may also be Sheetrock, depending on how many renovations have taken place. Many people believe these are assemblies found only on the east coast, but that is simply not true. They can be found all over the country.

Buildings that are wider than 20′ or stand-alone still use the basic Type III concepts of vertical masonry walls, embedded floor joists, elaborate front exteriors, and parapets. They have columns of cast iron or wood about every 15′ of width to support girder ends. These girder assemblies support the floor joists. The bearing exterior walls are six or more wythes thick.

FIGURE 6–23 These protrusions, or bump-outs, are common in Type III construction. *Courtesy of Michael Smith.*

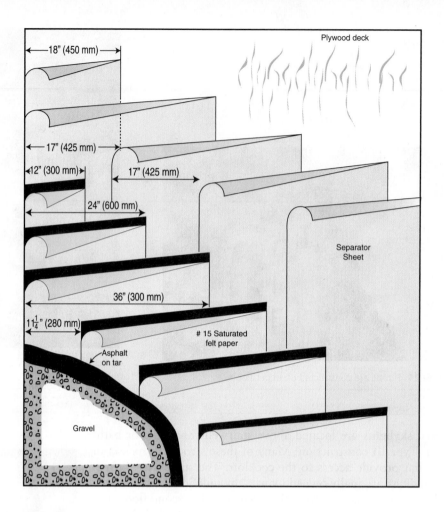

FIGURE 6–24 Construction features of built-up roofing.

Plywood deck

18" (450 mm)

17" (425 mm)

12" (300 mm)

17" (425 mm)

24" (600 mm)

Separator Sheet

36" (300 mm)

$1\frac{1}{4}$" (280 mm)

15 Saturated felt paper

Asphalt on tar

Gravel

If it appears as if an adjoining structure has been removed, firefighters should use extreme caution when conducting a fire attack. Type III structures built in a row were never intended to have a vacant lot between homes. If alleys or other roadways are present, the end units have thicker walls. If firefighters confront missing units, they are to perform only those operations that are absolutely necessary and attempt only confirmed rescues, They must be vigilant at all times because these buildings will fail quickly and catastrophically. They should be aware that hose streams can dislodge the lime mortar.

In the city of Philadelphia such catastrophic failures have occurred at least twice. In May 1993, three firefighters were trapped in a three-story Type III structure on American Street when the front wall collapsed suddenly.

It happened again on April 5, 1997, when a three-story building located at 718 N. 42nd Street suddenly collapsed, trapping a battalion chief and five firefighters. The firefighters had performed multiple rescues and had knocked down the fire on three floors. There were no firefighter fatalities, but this incident certainly underscores the dangers inherent in Type III buildings that have lost their lateral support partners.

FIGURE 6–25 Mansard roofs are part of the Federal style of architecture.

Type IV

Type IV construction is called mill, plank and beam, or heavy timber construction. This type of construction evolved because of the need for New England mills to house heavy manufacturing equipment. The interior structural members consist of massive wooden timbers. The minimum allowed size for major members of this assembly is 8″ × 8″. The girders supporting floor and roof assemblies run parallel to the ridge line. The assemblies intersect over columns approximately every 15′ to 20′. The ends of the wooden girders overlap using joints called half-laps. These use mortise and tenon connections or cast iron plates with cast iron bolts and nuts (see Figure 6–26). The girder joints can be supported by cast iron, wood, or masonry columns. Most often the columns on the lower floors will be masonry and the upper stories cast iron or wood.

The floor assemblies are major contributors to load support. They consist of 3″ to 8″ thick planks of wood connected by tongue-and-groove joints. These are 4″ to 10″ wide. They can be attached to floor joists by either wooden dowels or metal spikes. This is called *planking*. The floor joists are of 8″ or wider stock; depths of 20″ to 24″ are common. The joists are 6′ to 8′ apart.

The exterior is masonry or masonry with a stone foundation. The exterior walls have buttresses (massive masonry projections usually matching the lines of roof rafters) (see Figure 6–27) or *flying buttresses*. Flying buttresses are free-standing buttress assemblies connected to the main structure by an arch. The arch is not usually concealed (see Figure 6–28). These buttresses support the lateral force of the roof rafters as they attempt to flatten (see Figure 6–29). The buttresses also act to support the longitudinal

FIGURE 6–26 Typical timber connections.

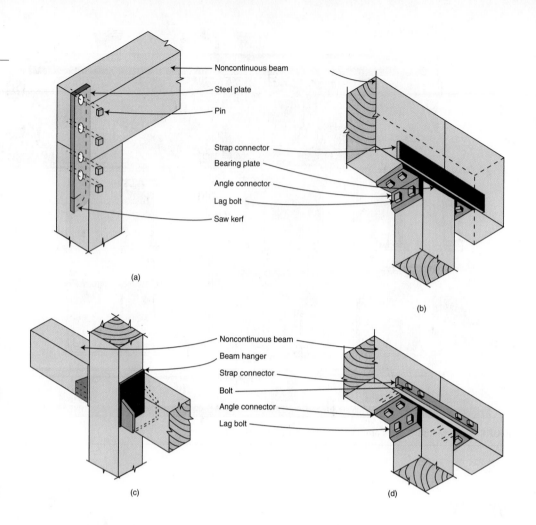

Noncontinuous beam
Steel plate
Pin

Strap connector
Bearing plate
Angle connector
Lag bolt
Saw kerf

(a)

(b)

Noncontinuous beam
Beam hanger
Strap connector
Bolt
Angle connector
Lag bolt

(c)

(d)

length of the masonry wall. Small structures may also be used to perform this function (see Figure 6–30). The interiors of Type IV constructed buildings contain projections of masonry called pilasters.

The interior spaces are usually divided into 50′ × 50′ areas defined by masonry walls. Arches or lintels are used at openings between spaces. Stairs can either be cast iron or wood and are quite steep to save space.

The roof rafters are usually the same dimensions as the floor joists. Most of these roof assemblies are pitched; the roof rafters can be trussed. The use of trusses dates back to the nineteenth century; the truss members are still quite substantial and have geometric shapes. The members use cast iron plates, cast iron rods with turnbuckles, split-ring connectors, cast iron rods with nuts and washers, or mortised-and-tenon wooden connectors (see Figures 6–31, 6–32, 6–33, and 6–34). The roof decking is also constructed of planking. The roof covering can be slate or metal sheeting (see Figure 6–35). Clearstory assemblies often include skylights along the ridge pole in.

All the attributes of this type of construction are also used in church construction, the major difference being the use of stained glass or painted windows and tall bell towers. The interiors are lavish, with ample uses of wood paneled and plaster walls.

These structures do not support fire readily, but once the fire gets headway large-caliber streams are required to extinguish it. As the fire progresses failures begin to occur throughout the building. Too often the result is total collapse of the structure (see Figure 6–36).

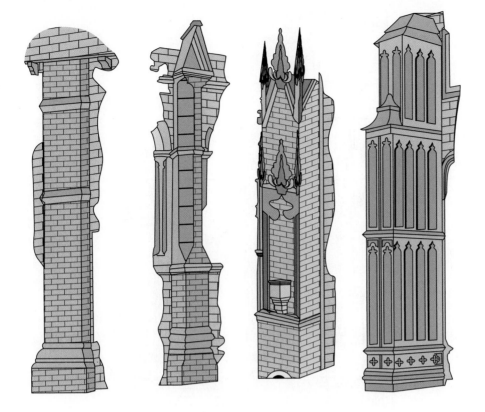

FIGURE
6–27 Buttresses.

Collapse zones are essential when combating fires in these assemblies. Recommendations for distance vary from one and a half to two times the height of the structure. Consider distance plus shielding. If a wall fails in a total collapse, the rubble can bounce and hurtle at firefighters with significant speed and force. A four pound brick hitting you at 80 mph will leave a mark. Get something solid between you and the structure or stay at the perimeter of the collapse zone.

Where codes allow, the modern approach to this type of construction uses microlam beams and other laminated major structural members. A microlam structural member uses smaller pieces of 2 × stock, such as 2 × 4, 6, 10, or 12, glued together (see Figure 6–37). Hence the term *glulam* (glued–laminated). Often the exterior of these assemblies are laminated with thin veneers that create the appearance of solid pieces. If curves are to be created the members are steamed as they are assembled (see Figure 6–38). These assemblies can fail quickly when exposed to water and fire; they will be discussed in more detail in Chapter 7.

Type V

Type V construction is entirely combustible and is the only type of construction that can be called fully involved. The major structural members are wood. The vertical wall supports can be 2 × 4s or 2 × 6s. The stud chases can be 12″, 16″, or 24″ on center. The floors and roof assemblies can be trussed. The wall sheathing or covering can be 1× wood planks, asphalt-impregnated fiberboard (commonly called Celotex), plywood, or, for newer construction, aluminized cardboard or OSB. OSB consists of wood chips, paraffin, and a glue binder cooked in water and processed into the same shape and form as multi-ply plywood.

FIGURE 6–28 Flying buttress.

 The use of SIPs (structural insulated panels) and QVE (quality valued engineered) framing practices can lead to even earlier structural failure than when using trusses alone.

The exterior coverings can be stucco over wire mesh, masonry, stone, aluminum or vinyl siding, and lightweight concrete simulated stone. This choice of siding does not change the construction type.

The roof covering can be asphalt or fiberglass shingles, terra cotta tile, standing seam metal, wooden cedar shingles (also called cedar shakes), or, if the roof configuration is flat, rolled membrane, cold mopped tar and asphalt paper, or built-up with aggregate. The use of manufactured slate, wood shakes, or terra cotta tile made from plastics is becoming more and more common.

The two most common forms of Type V construction found today are balloon framed and platform framed although post-and-beam are still present in older structures.

Post-and-beam construction came to the United States from Europe and was common in the nineteenth century in the homes of the wealthy. The large timbers used for major vertical and horizontal members were most likely hand hewn. The lack of skilled craftsmen eventually led to the decline of this type of construction.

FIGURE 6–29 Buttresses are an exterior indication of Type IV construction. *Courtesy of Michael Smith.*

FIGURE 6–30 Small structures take the place of buttresses. *Courtesy of Michael Smith.*

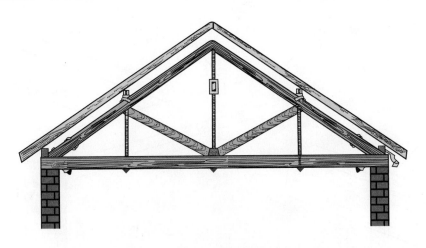

FIGURE 6–31 Early form of king-post truss.

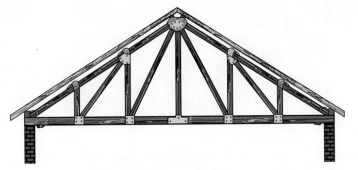

FIGURE 6–32 Early truss with cast iron plates at joints.

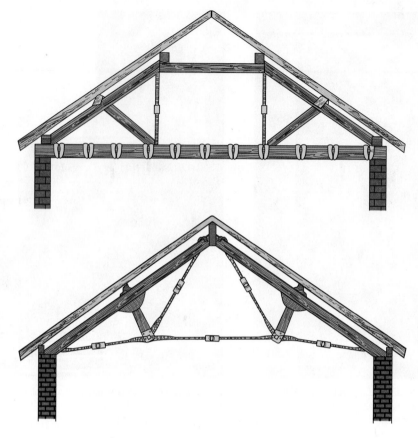

FIGURE 6–33 Open-post truss (early version).

FIGURE 6–34 Fink truss using cast iron turnbuckles.

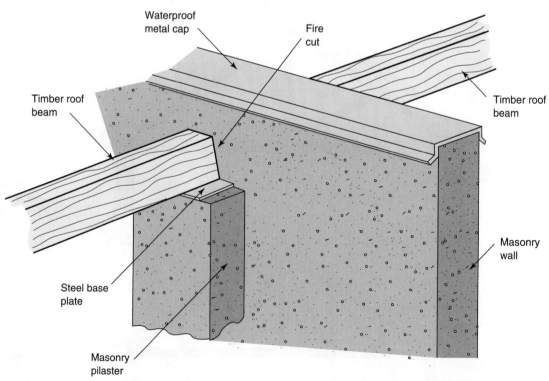

Waterproof metal cap

Fire cut

Timber roof beam

Timber roof beam

Masonry wall

Steel base plate

Masonry pilaster

FIGURE 6–35 Heavy timber frame to masonry wall connection. *Source: Andres, Cameron K.; Smith, Ronald C.,* Principles and Practices of Commercial Construction, *7th Edition, © 2004, p. 273. Reprinted by permission of Pearson Education, Inc., Upper Saddle River, NJ.*

FIGURE 6–36 Fires in churches can be dangerous. Respect collapse zones. *Courtesy of Michael Smith.*

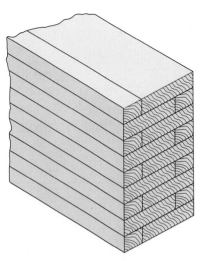

FIGURE 6–37 Glulam members consist of smaller dimension lumber glued together. *Source: Andres, Cameron K.; Smith, Ronald C., Principles and Practices of Commercial Construction, 7th Edition, © 2004, p. 280. Reprinted by permission of Pearson Education, Inc., Upper Saddle River, NJ.*

The evolution of Type V construction can be traced to the American Civil War and World War II. A major population shift occurred right after the Civil War. The southwest and western parts of the United States became more populated as veterans from both armies wanted to leave the devastation of the war behind. Gold was discovered in California during this period, creating the gold rush. Most of the towns constructed during this period were built in row configurations of balloon-frame construction, which consists of continuous vertical wooden supports. The roof assembly completes the system for structural stability. The floor system of joists and decking is added after the shell had been completed; this assembly is often in shear. The interior partitions and stairs are added last. It was during this period that the single-family balloon-frame homes began to appear. The roof configurations for these structures are pitched, either gable or hip. They often have dormers and almost all have porches. The major problem with this type of construction is the lack of horizontal stops in the exterior stud chases. The combustibility of the construction precludes the use of positive pressure ventilation (PPV) and allows fire and smoke to travel throughout the building. This type of framing was prevalent from the early nineteenth century until after World War II (see Figure 6–39).

After World War II the returning veterans, approximately twelve million strong, wanted their share of the American dream—the ability to own a home. The balloon-framing method required many 30′ to 40′ straight pieces of lumber for the exterior walls alone. This need far outstripped suppliers' abilities to secure, process, and deliver lumber. Another problem with balloon-framing was the size of the crew required for erection. The exterior walls were built

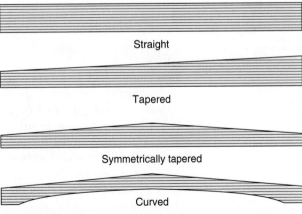

Straight

Tapered

Symmetrically tapered

Curved

FIGURE 6–38 Glulam shapes. *Source: Andres, Cameron K.; Smith, Ronald C., Principles and Practices of Commercial Construction, 7th Edition, © 2004, p. 281. Reprinted by permission of Pearson Education, Inc., Upper Saddle River, NJ.*

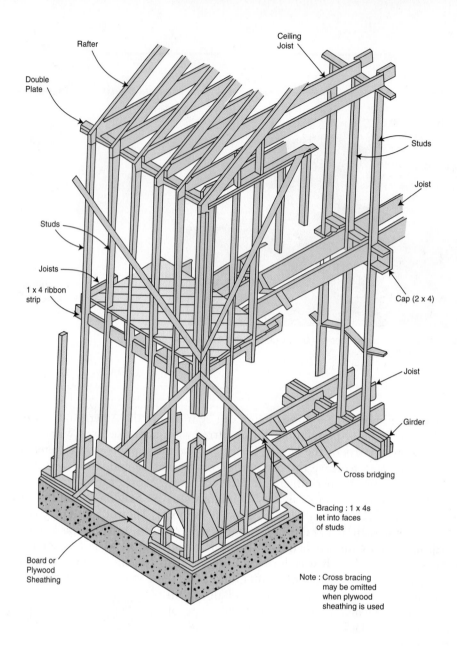

FIGURE 6–39 Balloon-framed structures cause problems relative to fire suppression.

Rafter

Ceiling Joist

Double Plate

Studs

Studs

Joist

Joists

1 x 4 ribbon strip

Cap (2 x 4)

Joist

Girder

Cross bridging

Bracing : 1 x 4s let into faces of studs

Board or Plywood Sheathing

Note : Cross bracing may be omitted when plywood sheathing is used

on the ground and raised to vertical. This required crews of 10–20 carpenters. The construction industry responded to the demand for new homes by developing platform framing (see Figure 6–40). *Note*: Plywood may be present in balloon-framed structures that have undergone renovation or have been rebuilt after a fire.

Platform framing immediately solved some of the problems associated with balloon framing and provided unforeseen benefits. The crew sizes could be much smaller, as few as five carpenters. The lengths of the vertical members were much shorter. The assembly process involved building a masonry foundation. The first-floor joists were installed directly on the foundation with a center girder traversing the length of the building. The vertical walls were erected with heights of 8' to 10'; subsequent story assemblies could be added. All main structural members were placed in compression. Trusses began to be used more frequently in floor and roof assemblies. See Chapter 7 for additional information on platform framing.

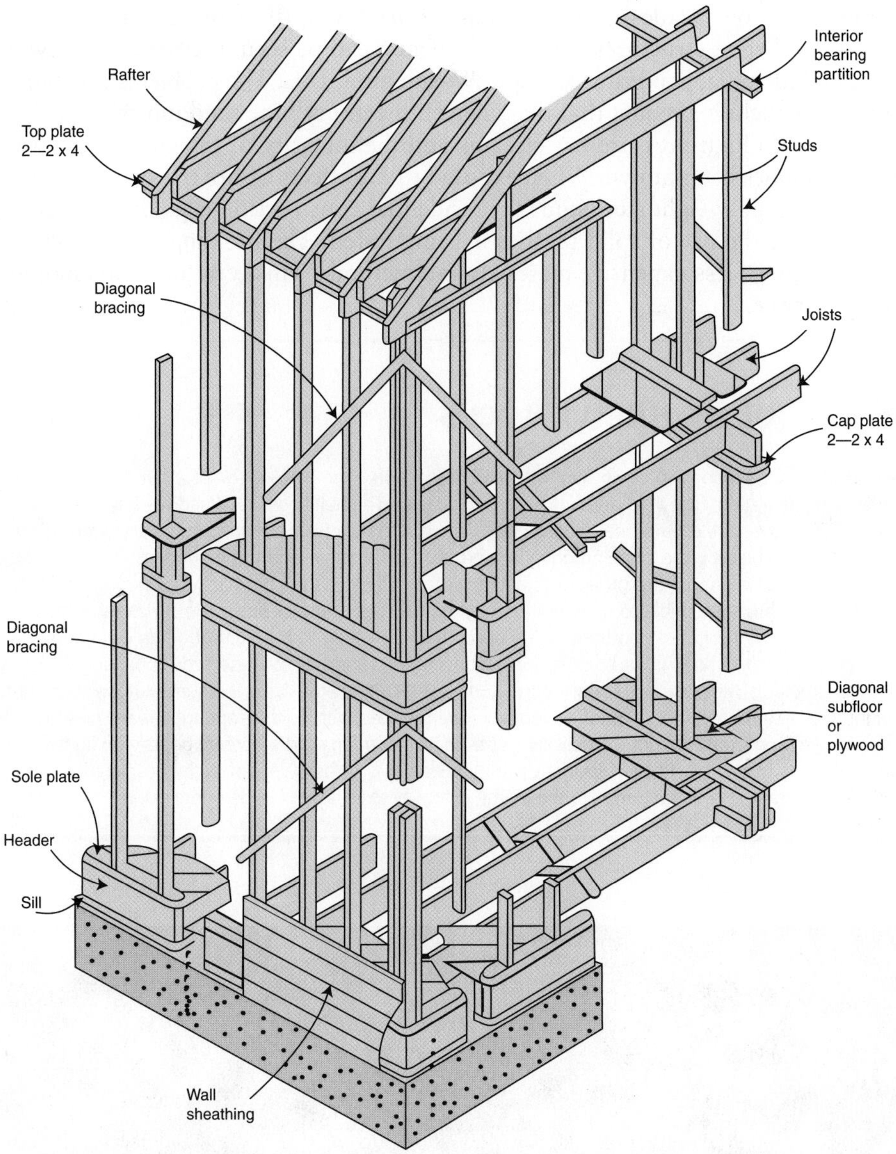

Rafter
Top plate
2—2 x 4
Interior
bearing
partition
Studs
Diagonal
bracing
Joists
Cap plate
2—2 x 4
Diagonal
bracing
Diagonal
subfloor
or
plywood
Sole plate
Header
Sill
Wall
sheathing

FIGURE 6–40 Platform framing creates compartmentalized residential fires.

TRUSSES

Early residential roof trusses were restricted to gable or hip configurations. Today any roof configuration can be trussed, although it is not a practical application for cones due to spacing requirements.

TGI floor and roof assemblies are currently eclipsing trusses. TGI is the common term for Truss-joist I-beams (TJI). A top and bottom chord of 2 × 2 or 2 × 4 with a plywood or OSB web is installed as joists or rafters. This assembly fails quickly, even faster than do trusses, when exposed to the effects of fire. Subcontractors can and do cut through the web to run their ductwork, pipes, and wiring, thereby further reducing the mass.

Hybrids

History tells us that most cities—precluding rebuilding necessitated by natural disasters and wars—are rebuilt every fifty years. Many cities attempt to retain a nineteenth-century flavor while employing twenty-first century methods. This is never more evident than when the fronts or façades of nineteenth-century Type III structures are attached to contemporary Type I structures. This phenomenon can be seen in cities such New York, Chicago, Philadelphia, San Francisco, and Washington, D.C. The problems this causes for the fire service include the fact that the nineteenth-century lime sand mortar can give way when exposed to high-pressure fire streams and the integrity of the connection between the two assemblies is unknown. When exposed to the effects of fire the structure may fail.

The use of lightweight steel studs for residential construction is gaining wider use; in some assemblies the use of solid plastics is being tested. The building industry continues to push for lighter, less expensive materials and methods. This certainly bears monitoring by the fire service.

SCHOOL OF HARD KNOCKS

Several years ago I made a training film for the FETN network. The piece revolved around the technique involved in integrating Type III façades into Type I structures. The buildings we filmed were all in Washington, D.C., but they could have been in any city. Some of the façades we highlighted were twelve stories in height. The projects were all being done in a professional manner and the workmanship was superb, however, I kept looking at the projects with the eyes of a fireground commander. The original mortar was used, which is susceptible to deterioration by the pressures of attack lines. The structural connection between the new and the old portions of the building is vulnerable to the heat of a fire. During my construction career I had been involved in major remodeling of hundreds of these Type III buildings. I knew that the structures were always brought up to code using state-of-the-art techniques. But we used lightweight truss systems for floors. And the windows were replaced with double- and triple-pane glass, which contains heat and smoke longer, complicating size-up. We also replaced solid core doors with lightweight pre-hung units that will fail faster. Incident commanders need to be aware of such considerations when called to emergencies at structures like these. The best way to be safe is to be informed.

FIRST RESPONSE

Residential Structures

Residential structures represent the bulk of the fires involving fires for the fire service. Residential fires are referred to as bread and butter fires. These occupancies are normally Type V. They may be detached (stand-alone), duplex (two homes side by side), or row (many homes constructed together). The newer versions of row homes use Sheetrock for the party wall. They can be one or more stories in height. In some areas of the country two or three residences can be built by stacking one on top of the other, forming double and triple deckers. Many will have fireplaces or free-standing wood stoves. They may be sprinklered; if they are, sprinklers will only be in areas allowing for escape, such as a hallway. The windows in modern homes will usually be double- or triple-pane. Older homes may contain double- or triple-pane replacement windows, but the original windows will be single-pane plus storm windows with screens.

The framing may be conventional, using sawn material and standard stud and rafter spacing of 12", 16", or 24" on center; or it may be a combination of conventional wall framing and trussed floors and roof. The framing can utilize SIPs or QVE framing techniques. The framing members can also be lightweight C-shaped steel units. The exterior can be clad in masonry, vinyl, aluminum, wood, stucco, imitation stone, or a combination of these materials. The roof can be covered with any number of options, including metal standing seam, fiberglass shingles, asphalt shingles, terra cotta tile, or imitation manufactured shingles replicating wood shake, terra cotta tile, or slate. Some codes allow for minimum spacing between detached structures.

Some considerations for strategies and tactics are listed below:

- Most, if not all, departments are equipped to handle one or two rooms engulfed in fire in these occupancies.
- Protect all searches and rescue attempts with hoseline coverage as soon as possible.
- Incident commanders and company officers need to know conditions in the rear and basement before developing tactics and strategies.
- Consider alternative methods for vertical ventilation if the roof is suspected of being trussed and fire is impinging on the top floor.
- Ensure that adequate ladders for firefighter evacuation are provided when units.are operating above the first floor.
- As firefighters enter the structure, test the floor for heat as it may indicate that the fire is below them.
- Ensure that accountability is maintained during the fire fight.

ON SCENE

Fire units have been operating at the scene of a fire in a six-story government building of Type I construction. The roof, which appears to be giving off smoke, is constructed of concrete. There is a roof assembly on top of the concrete that consists of wooden and steel rafters with wooden sills; the insulation between the sills and the concrete is burning. The incident commander orders you to make trench cut in the roof deck. There has been a partial collapse and you can see a cross section of the building at a point away from where the chief wants the trench cut. You have three engines and two truck companies under your direction.

1. Would you attempt a trench cut in a Type I building? Why or why not?
2. What safety concerns would you have?
3. What other method could you employ?

CHAPTER REVIEW

Summary

Construction is categorized according to type; the five major types have recognizable characteristics. They also have varying levels of combustibility, from fire-resistive as in Type I to entirely combustible as in Type V. It is important for incident commanders and firefighters alike to understand the basic elements of each type, both for the purposes of safety when they are called to size-up as well as in order to fight fires at these structures.

Construction features for each type are the same no matter where the buildings are located in the country. Problematic features have been highlighted in this chapter, especially the trend toward updating one type of structure by adding the façade of another type of construction. Subjected to fire, a structure can fail due to the fact that older building materials were retained when the building was remodeled. Use caution when assuming that all types will react in the same way in which they have in your past experience.

By familiarizing yourself with each type of construction, you will create a safer firefighting experience for everyone involved.

Review Questions

1. Discuss the principal difference between Type I and Type II construction.
2. What does *fire rated* mean?
3. How is performance measured for fire rating?
4. Discuss how concrete can be used for floor assemblies in multistory (high-rise) structures.
5. Discuss how specification versus performance codes will affect structural integrity during fire or other disasters.
6. Why would Type IV structures cause problems for fire personnel?
7. What are collapse zones?

Suggested Reading

Andres, J. 1998. *Principles and Practices of Heavy Construction.* Upper Saddle River, NJ: Pearson Education.

Dechiara, J. 1980. *Handbook of Architectural Details for Commercial Buildings.* New York: McGraw-Hill.

Duval, Robert J. 2006. *NFPA Case Study: Nightclub Fires.* Quincy, MA: National Fire Protection Association.

Honkala, T. 1994. *Carpentry and Light Construction.* Upper Saddle River, NJ: Pearson Education.

Love, T. 1970. *Construction Manual: Rough Carpentry.* Solana Beach, CA: Craftsman Book Company.

NFPA 220. 1999. *Types of Buildings.* Quincy, MA: National Fire Protection Association.

Smith, R. 2004. *Principles and Practices of Commercial Construction.* Upper Saddle River, NJ: Pearson Education.

Trusses and Other Manufactured Assemblies

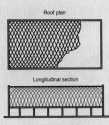

KEY TERMS

bottom chord, *p. 166*
glulam beams, *p. 177*

gussets, *p. 166*
top chord, *p. 166*

web, *p. 166*

OBJECTIVES

After reading this chapter, you should be able to:

- Identify all of the major components of a truss.
- Recognize truss construction.
- Understand the forces acting upon trusses and their dangers.

Resource**Central**

For additional review and practice tests, visit **www.bradybooks.com** and click on Resource Central to access book-specific resources for this text! To access Resource Central, follow directions on the Student Access Card provided with this text. If there is no card, go to **www.bradybooks.com** and follow the Resource Central link to Buy Access from there.

Introduction

Trusses are used in many types of structures today, not just in tract housing subdivisions. You can find trusses in multistory multiple-occupancy buildings, offices, warehouses, and high-dollar as well as low-income houses.

Trusses have been in use for over 500 years. Some of the earliest forms of trussing can be found in the railroad systems in the United States where they were used to construct bridges by connecting the members using cast iron tie rods or cast iron plates; early uses can also be found in cathedrals in Europe. Many of these earlier designs incorporated wood as the main resource. Wood was replaced by cast iron in the mid-nineteenth

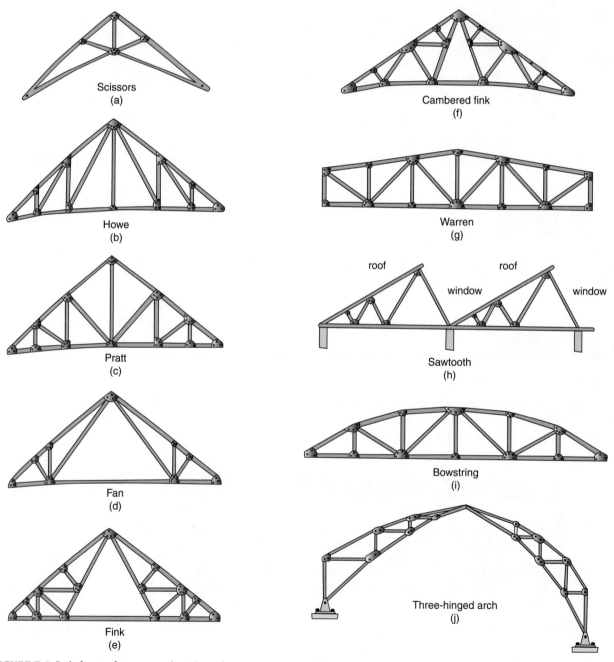

FIGURE 7-1 Early forms of trusses used cast iron plates to connect members.

century; steel was used in the late nineteenth century and early twentieth century. Many early trusses used wood with cast iron plates for the assembly (see Figure 7-1). The Warren truss was patented in 1848 by its designers, James Warren and Willoughby Theobald Monzani. The Pratt truss was patented in 1844 by Caleb Pratt and his son Thomas Pratt, two Boston railway engineers (see Figure 7-1).

The hammerbeam truss was used throughout Europe when an ornate and elaborate roof rafter was needed to span large distances. The hammerbeam truss is still used today in many public buildings and churches (see Figure 7-2). Many of the trusses used in today's construction are descendants of these earlier truss forms.

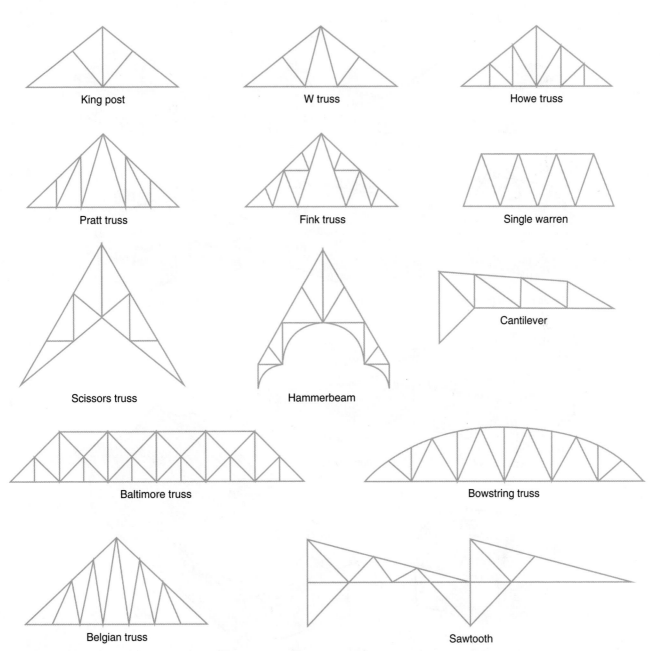

FIGURE 7-2 Common truss configurations. *Source: Andres, Cameron K.; Smith, Ronald C.,* Principles and Practices of Commercial Construction, *7th Edition, © 2004, p. 407. Reprinted by permission of Pearson Education, Inc., Upper Saddle River, NJ.*

After World War II the need for housing outstripped the lumber industry's ability to supply the wood needed. And because conventionally framed buildings required many craftsmen to construct them, technology was created to address the lack of craftsmen. Engineers were able to calculate the loads delivered to the various framing members. They then designed geometric patterns that addressed the physics of constructing a wooden or steel member that could support the same load with smaller component parts. One serendipitous result was that larger open areas could be created within a structure because fewer internal framing supports were needed. The use of trusses has burgeoned since then. Just about every configuration of roofs can be trussed (see Figure 7-3).

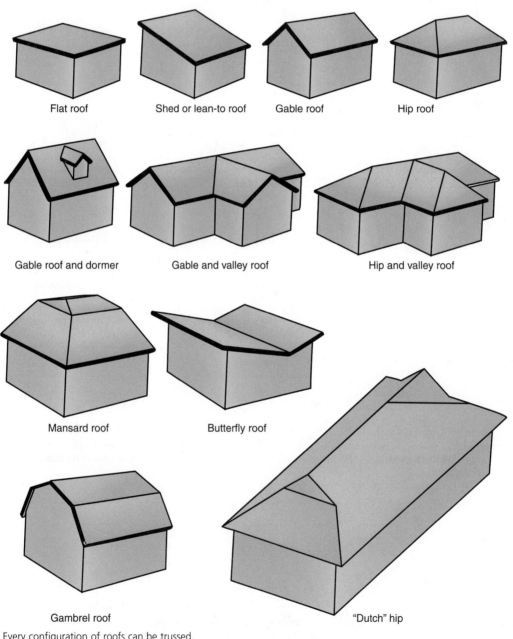

FIGURE 7-3 Every configuration of roofs can be trussed.

The savings in dollars for the end user have driven the truss industry to continue to push the mass versus support equation. It would be a safe bet for a firefighter to assume that any residential and most commercial buildings constructed in the last fifty years will contain trussed components.

Many disasters involving trusses and firefighters, including the Hackensack car dealership in Hackensack, New Jersey, a church in Memphis, Tennessee, and an auto parts supplier in Chesapeake, Virginia. There have also been many incidents at residential buildings that have had small to total failures. Even with this history well documented and with many hours of related discussion, firefighters continue to operate in, under, and around trusses while seemingly oblivious to the dangers.

FIREHOUSE DISCUSSION

The Waldbaums Supermarket located at Ocean Avenue and Avenue Y in the Sheepshead Bay section of Brooklyn, New York was in the process of being extensively renovated. On August 2, 1978 a fire was reported in the mezzanine area at 0815 hours. The fire department responded with three engine companies and two ladder/truck companies. A battalion chief also responded. The ladder companies proceeded to ladder the roof and began to ventilate using saws and other hand tools. To ventilate means to open holes in a roof or floor above a fire to release the heat, gases, and pressure from the area below. The standard operating procedures for the FDNY call for ventilation to occur in this scenario.

Without warning the roof collapsed, sending twelve of the twenty firefighters operating on the roof into the inferno below. This occurred approximately thirty-two minutes into the incident. Six of the twelve escaped; the other six were killed.

The Hackensack Ford fire occurred in Hackensack, New Jersey on July 1, 1988. The fire occurred at approximately 3:00 PM. Workers in the auto dealership and repair center reported that the fire appeared in the attic space above the repair portion of the building. Ladder companies responding to the first alarm began ventilating the roof while engine companies attempted to attack the fire from the interior. Approximately thirty minutes after the arrival of the first units the roof came down, trapping the five firefighters who were manning hose lines. Despite valiant efforts the five firefighters were killed.

Both of these structures contained bowstring trusses. These assemblies have been blamed for the lost firefighters, but as a radio commentator once said, "and now for the rest of the story."

The Waldbaums Supermarket had been undergoing a significant renovation. The original roof was constructed of bowstring trusses and was covered with a composite roof of paper, tar, and stone chips. At the time of the fire, however, the roof was covered over with a rain roof. This prevented the firefighters from realizing that the building had a bowstring truss roof assembly under the flat roof rain roof. A rain roof is constructed by erecting knee walls and installing a truss, usually parallel chord, over the existing roof structure. A new roof covering is applied and the space above the original roof is left intact. The framed walls are then covered over with another treatment, which is usually different from the original. The treatment might be masonry, which will appear different. It could be a stucco look using either parging or drivit. Or it could be some other exterior treatment. The Waldbaums building appears to have been covered with stucco, resulting in a false impression being gained by the firefighters because the intact original assemblies hid the effects of the fire and allowed it to compromise the entire roof structure.

The Hackensack Ford structure had also undergone significant renovations. The original support for the ends of the bowstring trusses had been masonry walls. These trusses spanned 78' and were spaced 16' apart. The trusses consisted of segmented pieces of 3" × 10" sawn wood connected by spliced joints. The joints were bolted together. The new supports consisted of steel columns with steel beams bolted to the trusses. A 20' × 175' service area with a flat roof was then added onto the building. The ceilings in the affected area were 20' high and covered with cement plaster and wood lathe. The attic area was used as storage for a variety of automobile parts including car engines. The attic storage area comprised 7,800 square feet. The fire began in the attic area and attacked the segmented pieces of the truss. The building, which was 100' wide, was 224' on the north side and 175' on the south side.

During my forty years in the fire service there has been an ongoing debate regarding trusses and their connection to firefighter fatalities. Trusses are an engineering marvel. They have provided the building industry with a valuable tool with which to support loads over wide spaces without intermediate support system components. But: trusses are part of a total system that absolutely has to remain intact completely, without exception. Documentation from NIOSH, Firefighter Close Calls, and others indicate that if flame and heat are impinging on truss assemblies, those assemblies are going to fail. Anyone on top of or under this failure would not be going home that day or night. The resistance to applying water in these structures by the utilization of master streams is rooted in the philosophy that an "aggressive interior attack is always best."

There remains an axiom that if you apply water fast enough, there won't be any problems, but the inherent weaknesses of trussed assemblies continue to accumulate its victims every day.

Research the number of firefighter fatalities associated with truss failures that have occurred over the past five years.

- Determine whether your community has a code requiring trussed buildings or buildings where a rain roof has been installed to be marked.
- Check the training opportunities in your area and all surrounding jurisdictions to determine how much training is dedicated to truss awareness.
- Discuss how much time should be expended before an offensive attack goes defensive in commercial structures.

Truss Basics

Wooden and steel trusses are similar in that they both support maximum loads using the minimum mass of materials. This is explained by the differences between solid sawn lumber, steel, and trusses. The trusses' depth, or the distance between the top and bottom chords, is what measures its efficiency. Adding more members to the trusses' web and increasing the size of the chord materials allow the truss to carry the necessary load. To span the same distance and support the same load without intermediate supports, the solid sawn wood materials would be heavier and much less cost effective. The exact same rules apply to steel trusses.

The loads are carried by geometrically-shaped forms. The triangle is used most often because of its structural stability. A triangle with sides of fixed lengths is the simplest geometric shape that does not change shape under load conditions. The load, span, and spacing are factors that an engineer uses to determine truss configuration and the size of the members.

WOODEN TRUSSES

First, let's look at the truss itself. Wooden trusses are constructed of three major components: the upper member or **top chord**, the lower member or **bottom chord**, and the intermediate members or **web**. The size of the members is usually 2 × 4 or 2 × 6 depending on the total span to be bridged or the anticipated load. The components are joined together not by nails, but by lightweight pieces of metal, called **gussets**, which are stamped to create small projecting points that penetrate the wood to the depth of only about ¼"—and this is if the punch machine is at full strength and applied properly (see Figure 7-4).

Newer trusses no longer use gussets or sawn lumber. They are assembled using multi-ply wood (similar to plywood in composition and design) and are connected by the use of a groove cut into the top and bottom chords into which the web members are inserted. The web members are then glued into place using an isocyanate resin. When the chord members are not long enough, the manufacturer will splice pieces together using a field splice (see Figure 7-5).

top chord ■ The top member of a steel or wooden truss. It is also the top member of a steel girder.

bottom chord ■ The lowest horizontal member of a truss; the term applies to both wood and steel trusses.

web ■ In steel construction, the wide part of a beam between flanges; a diagonal support member in a truss.

gussets ■ Reinforcing members at the intersection of parts needed to complete an assembly.

The height of the web members is dictated by the span to be bridged and by the spacing of the trusses. Once the span is established the truss manufacturer sets up a jig, a table with various attachments designed to hold pieces in place until fixed, to receive the wooden members needed to fabricate the truss assembly. The key to remember is that there will be a joint at the center of each truss. This joint will occur in the middle of a space where the span is at its most critical moment, resulting in a span that is extremely prone to failure should it become impinged by flame or heat. There are two ways to connect the joint: a gusset comprised of lightweight metal punched to form the many little ferrules or spikes needed to connect with the wood or chord members grooved with the web members inserted into it. The total length of the truss is factored in and each side of the length is constructed from the center out to the ends. This is usually done in 10′ to 12′ increments with each section being joined using gussets or glue. The vulnerability of all wooden trusses lies not with the wooden members but at each joint. The connections will fail before the members.

There are three basic shapes for trusses: pitched, flat, and arched. The pitched, or common, truss can be recognized by its triangular shape. This truss is most often used for roofs. Some common examples of this truss are the king post truss, the scissor truss, and the Fink truss.

The king post truss is a pitched truss with a single vertical member in the web. This truss is used primarily for roof construction. It can span between 24′ and 36′ based on whether it will be affected by snow load. For heavy timber applications, the spacing between trusses is usually kept to 16′. For lighter weight residential applications, the spacing is usually kept to 19″ to 22″ (see Figure 7-6).

The queen post truss is similar to a king post truss, but it will have two vertical members in the web (see Figure 7-7). This truss is often used to create areas inside the truss space for storage or the HVAC system's air handling unit(s). Anticipate increased loads on the truss and be cautious.

FIGURE 7-4 Truss plate connection occurs at all intersecting points. *Source: www.shutterstock.com. Copyright: David Huntley.*

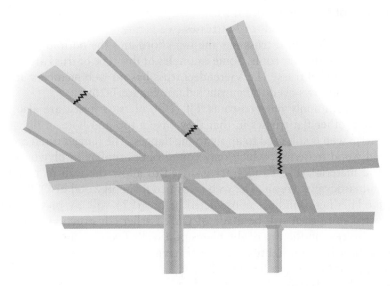

FIGURE 7-5 Field splice.

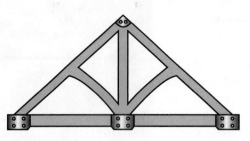

FIGURE 7-6 King post truss.

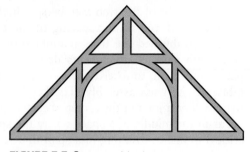

FIGURE 7-7 Queen post truss.

FIGURE 7-8 Scissor truss with cable.

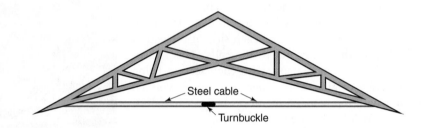

The Fink truss is used mainly for large spans over 50'. It can be recognized by the three isosceles triangles that form the web. One triangle is in the center, with its base formed by the bottom chord; the other two triangles form their base along the top chord [see Figure 7-1(f)].

The scissor truss is a pitched truss used when a vaulted ceiling or peaked ceiling is called for. With this truss much of the load will be transferred to the supporting bearing walls, so restraints on these walls will be typical. It is used exclusively for residential use, but if constructed of heavier timber materials, it could be used for larger commercial projects. Another method for increasing the span is to use a cable and turnbuckle assembly attached to the bottoms of the scissor (see Figure 7-8).

The hammerbeam truss is a pitched truss widely used today in churches or other public buildings requiring a regal appearance for their ceiling area. It is an elaborate truss system and when ornate wood trim is installed, it is unrecognizable as a trussed assembly. This truss most often requires buttresses on the exterior of the building, however, a modified version of the hammerbeam truss incorporates a vertical king post in the center of the truss, alleviating the need for buttresses (see Figure 7-9).

The flat or parallel chord truss has the top chord and the bottom chord in parallel and is used for floors and flat roofs. Some examples of this type of truss are Town's lattice truss and the Vierendeel truss. The Vierendeel truss has no web members and so is used when doors and windows need to be installed (see Figure 7-10).

The I-joists, commonly referred to as TJI, TGI, or truss joists, are also flat trusses. They differ from other flat trusses in that they have a top and bottom flange instead of chords. The term web is still used to apply to the flanges. The flanges often are made from manufactured wood materials, which may contain plies similar to plywood or be made of long strand material. The flanges are grooved to allow the web to be inserted. The grooves contain isocyanate resin. The web is formed from plywood or OSB depending on the age of the building. Buildings constructed before 1975 will use plywood, and buildings constructed since then will be made of OSB. Earlier construction techniques called for the I-joists to rest on the top of the plates of a wall. Today you should expect to find the I-joists suspended from lightweight metal hangers, in shear in the bearing walls, or from manufactured beams (see Figure 7-11).

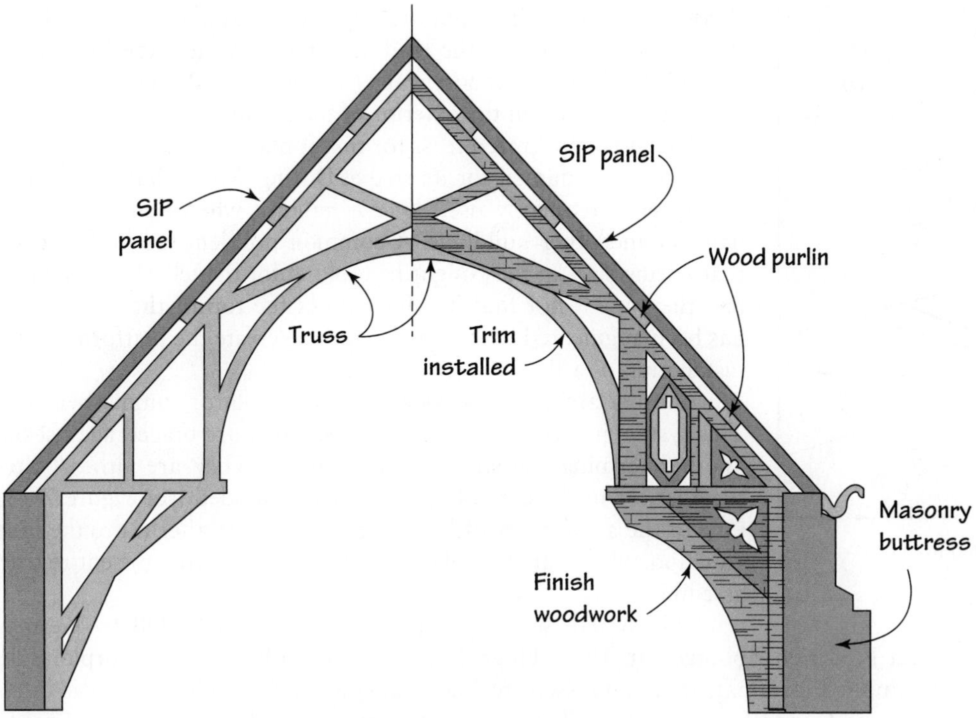

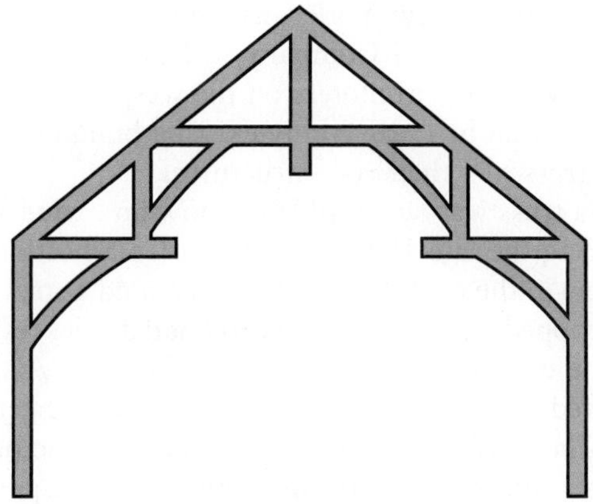

FIGURE 7-9
Hammerbeam and
modified hammerbeam
trusses.

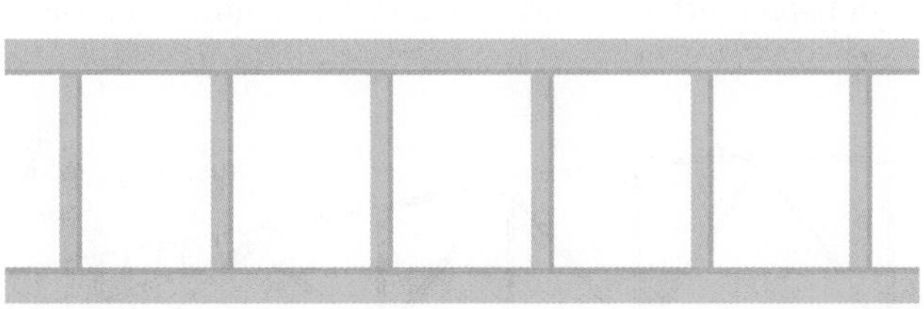

SIP panel

SIP panel

Wood purlin

SIP
panel

Truss

Trim
installed

Masonry
buttress

Finish
woodwork

FIGURE 7-10 Vierendeel
truss.

The arched-type truss was used from the late nineteenth century until the mid 1960s and mainly in commercial buildings. This truss system allowed for wide-open floor spaces with few or no columns. While flat and pitched trusses usually only have two bearing walls, the arched-type system uses all four walls to create the arched shape. This type of truss is often called a heavy timber truss because it utilizes larger members to form the arched top chord, the web members, and the flat bottom chord. While most trusses transfer loads vertically, this truss, because of the arch, tends to transfer the load laterally. This

Chapter 7 Trusses and Other Manufactured Assemblies   **169**

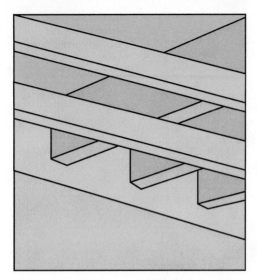

FIGURE 7-11 I-joists suspended from beam with hangers.

truss system often incorporates hip trusses at the ends to absorb the stresses imposed by the load, and then transfers the load to the ground through the support walls (see Figure 7-12). This truss often requires buttresses on the exterior of the building.

The most recognized truss for this application is the arched bowstring truss, so named for its shape. During World War II this truss system was frequently used by the military when building aircraft hangars and other buildings. A common problem for firefighters is that during size-up they might be looking for the telltale curved roof and they might not find it. Many times the top of the arched roof has been augmented with vertical studs to create the platform for the installation of a flat roof (see Figure 7-13).

Another arched truss system is the lamella or summerbell roof. This system incorporates short members that are braced against one another forming the shape of a diamond. They are either bolted together, or in the case of steel, bolted or welded (see Figure 7-14). One of the advantages of lamella trusses is that after a fire the individual members can be replaced without endangering the entire truss assembly.

This system was developed in Germany in 1908 by German architect Frederich Zollinger. In 1924, Hugo Junkers, soon to be a famous airplane designer, applied for a patent for a pressed steel lamella system. The technology was introduced in the United States in 1925. Two separate entities owned the proprietary rights. The Lamella Roof Syndicate in New York City and St. Louis were the east coast and midwest promoters. The Summerbell Company in Los Angeles and Timber Structures, Inc. of Portland, Oregon were the promoters on the west coast. The lamella truss gained traction in southern California in 1926. Many existing buildings in Long Beach and San Francisco have lamella trusses in their roof structures.

The wooden lamella truss was susceptible to snow and wind loads. The steel lamella truss came into use during the late 1930s. The Coca Cola bottling plant in Los Angeles has a lamella truss system in the roof. World War II created a steel shortage at which time the steel lamella truss stopped being used and wood had a brief resurgence, but the earlier vulnerabilities led to the overall demise of the lamella truss. Also adding to its demise were the use of laminated wood products that allowed for longer spans and the advent of the geodesic dome. The geodesic dome is based on the same principles as the lamella truss: isosceles triangles joined together to form the domed shape.

Lamella trusses made another appearance during the 1960s; wood was used in residential and light commercial projects and steel and aluminum were used in larger projects.

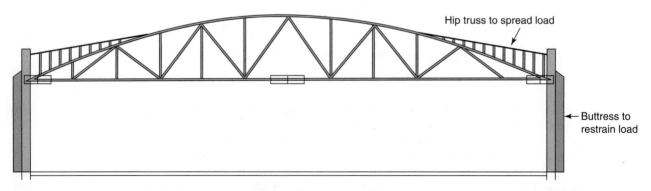

FIGURE 7-12 Bowstring truss with hip extensions for load transfer.

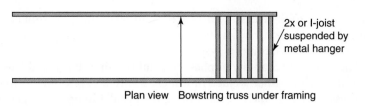

2x or I-joist
suspended by
metal hanger

Plan view   Bowstring truss under framing

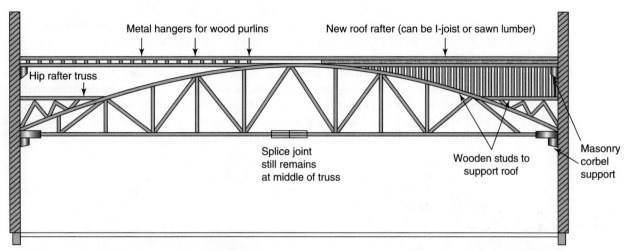

Metal hangers for wood purlins   New roof rafter (can be I-joist or sawn lumber)

Hip rafter truss

Splice joint
still remains
at middle of truss

Wooden studs to
support roof

Masonry
corbel
support

FIGURE 7-13 Framing for flat roof on top of bowstring truss.

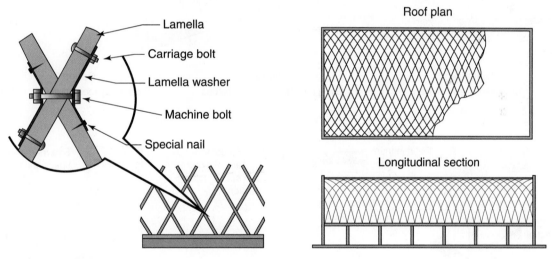

Lamella

Carriage bolt

Lamella washer

Machine bolt

Special nail

Roof plan

Longitudinal section

FIGURE 7-14 Lamella truss.

The Astrodome in Houston, Texas and the Superdome in New Orleans, Louisiana are both constructed with steel lamella trusses.

Wooden trusses are an excellent means of supporting weight under compression, but they are absolutely terrible in tension. This is similar to the limitations of concrete and is caused by the inherent limitations of mass versus geometry. Trusses are used to cover maximum lengths with maximum loads with minimum mass. The time when a truss is most vulnerable to failure is when it's going up or when it's coming down. If you have ever watched trusses being installed you have seen how unsteady they are before they are tied in by the roof covering. When trusses are first installed, lengths of wood are nailed perpendicular to the trusses. These are left in place until the roof covering is installed (see Figure 7-15). It is only when the roof decking is laid and connected perpendicularly to the trusses that it becomes a system.

The same vulnerability applies to metal trussing. For metal trusses, the long rod visible in the web is either welded or attached with a clip to the bottom chord.

Trusses are extremely fragile during the installation phase. They will roll or twist until they have been attached to the supporting wall. The only difference between metal and wooden trusses in regards to this vulnerability is the attachment method. Wooden trusses are nailed and metal trusses are welded or bolted using high strength bolts (see Figure 7-16).

Builders can use thinner plywood on the roof if they use H clips. The clips are made from aluminum or lightweight metal. By their name you can guess they are shaped like the letter H. Most codes require ½" plywood with two foot centers on the truss spacing. But if builders use H clips, they can go to ⅜" plywood or OSB board. The clips will fail under heat conditions. They will also fail under the weight of a firefighter's boot. Do not be lulled into thinking the roof is safer because the truss in place.

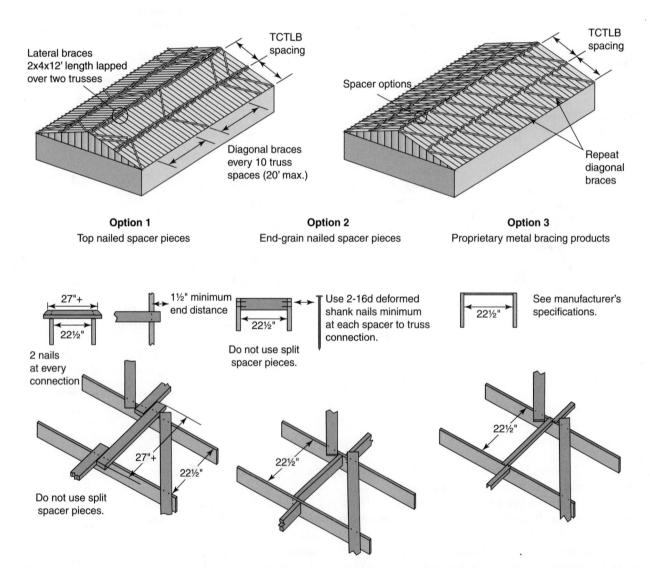

FIGURE 7-15 Bracing directions for trusses. *Source: Graphics and table reprinted with permission of WTCA-Representing the Structural Building Components Industry and the Truss Plate Institute. (For more information, visit www.sbcindustry.com and www.tpinst.org. For more information on fire service resources, visit www.sbcindustry.com/firepro.php.)*

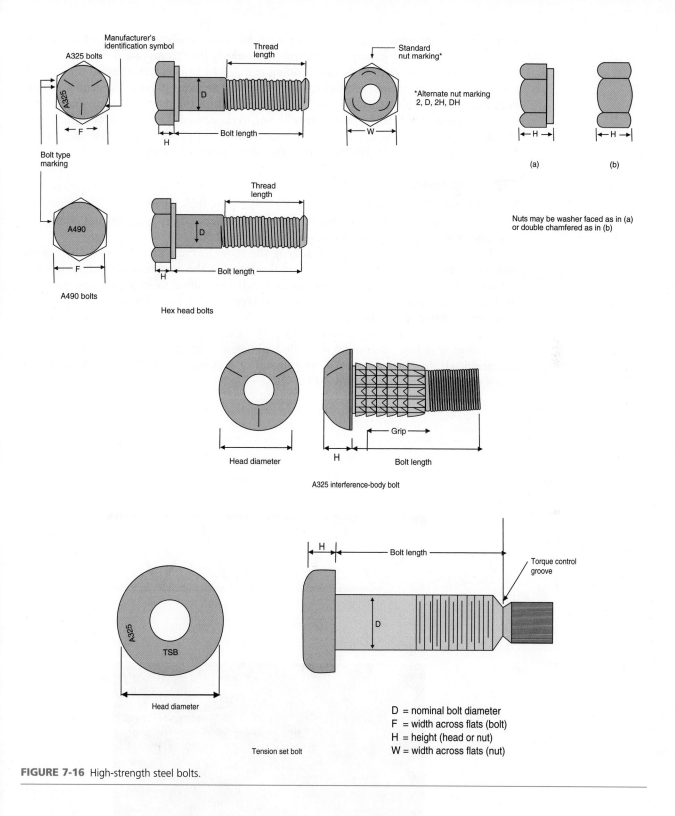

FIGURE 7-16 High-strength steel bolts.

If you do have an incident at which you suspect the construction is trussed or better yet, you have catalogued the building from its beginnings and know it is trussed, then you have to take extremely measured actions. First, if the fire is suspected of attacking the floors or roof, you absolutely can't be standing on or under this area. You need to ventilate from the ends of the gable roof or nonbearing walls, or if the roof is another configuration (hip, for example), the venting needs to be accomplished without getting onto the roof. If you have to be inside doing search and rescue, stay close to the bearing walls. If the failure occurs, remember two things: the weakest point is the center and the ends most likely have at least 8′ to 10′ of wood before the connecting gusset. Use the reach of your attack line so you can avoid being in the center. This is true for either wood or metal. For metal, it is key to get water onto the metal to stop the expansion process. For wood, it is key to remain away from the center of the structure.

STEEL TRUSSES

The steel bar joist is constructed using the basic principles of its wooden counterpart. There are a top chord, bottom chord, and web members. Many types of roof configuration can be achieved using steel trusses (see Figure 7-17).

The size of the truss is calculated using the same formula as for wood. The biggest differences are the raw materials used and the fact that the members are welded to form the connecting points. A metal truss can be welded or bolted to its supporting member, or it can be set into pockets of masonry. This type of truss has a flaw: when exposed to heat and/or flame it will begin to expand at which point it will lose its tensile strength and its ability to carry weight or support itself (see Figure 7-18). The movement of the truss causes it to lose its contact with the building and thereby fail before the failure of the steel itself. Steel trusses can support metal Q decking with lightweight concrete for floors or roofs above (see Figure 7-19).

FIGURE 7-17 Open-web steel joist.

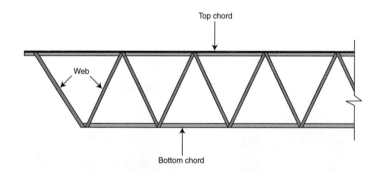

FIGURE 7-18 Steel deck on open web steel joists.
Source: www.shutterstock.com. Copyright: Alexey Fursov.

FIGURE 7-19 Steel trusses resting on support points. Failure occurs as the truss moves. *Source: www.shutterstock.com. Copyright: Jim Parkin.*

TRUSS FAILURES

Wooden and metal trusses fail differently, but they will both fail under fire conditions.

Wooden trusses have connecter issues. At every intersection of assembly the attachment point or connection will produce the fatal flaw (see Figure 7-20). The connections will fail before the wood burns through.

The heavier trusses (heavy timber trusses) fail also. Even though they are manufactured from larger stock, usually a minimum of 3″ × 10″, they are still connected by steel gussets, steel cables with turnbuckles, or cast iron rods with nuts and washers. When connected with gussets, the gussets are a minimum of ¼″ thick steel and utilize bolts and nuts to connect the pieces together. The steel cables and cast iron rods are usually ¾″ to 1⅛″ in diameter. The heavy timber trusses are constructed with segments of sawn lumber or larger pieces of laminated manufactured lumber. They have the exact benefits and drawbacks as their smaller relatives. They are used to span the greatest distance while supporting the greatest loads with the smallest sized material, all of which are arranged to share the load. These behemoths are excellent at combating gravity, but expose them to flame and heat for a long enough period and they will fail catastrophically.

When wooden trusses fail they will do so in a pancake collapse, V-shaped collapse, or lean-to collapse. Succeeding collapses of adjoining trusses will most certainly occur. See Chapter 9 for additional information.

Metal trusses are either welded or bolted together (see Figure 7-6). When manufactured correctly these assemblies are very strong. Metal truss ends will rest in pockets created in masonry walls or will be welded or bolted to metal attachment plates anchored to the masonry in shear. At intermediate connections the metal trusses will rest on masonry or metal columns.

When the temperature of the steel approaches 1,100°F, the steel tends to elongate. This causes it to push through the masonry laterally at the ends thereby causing the intermediate columns to move from axial loading to eccentric; this results in their failure. As the steel continues to be heated, it begins to change internally; at approximately 1,500°F it loses tensile strength and starts to droop, and finally at 2,000°F the steel fails completely in a pancake, V, or lean-to collapse. The load shifting causes other parts of the structure to fail as they become overloaded.

Some fire departments have had success in having buildings constructed with trusses (either wood or metal) to be marked with some form of icon, however, this is not easily accomplished. The necessary legislation might result in the occupants of these buildings

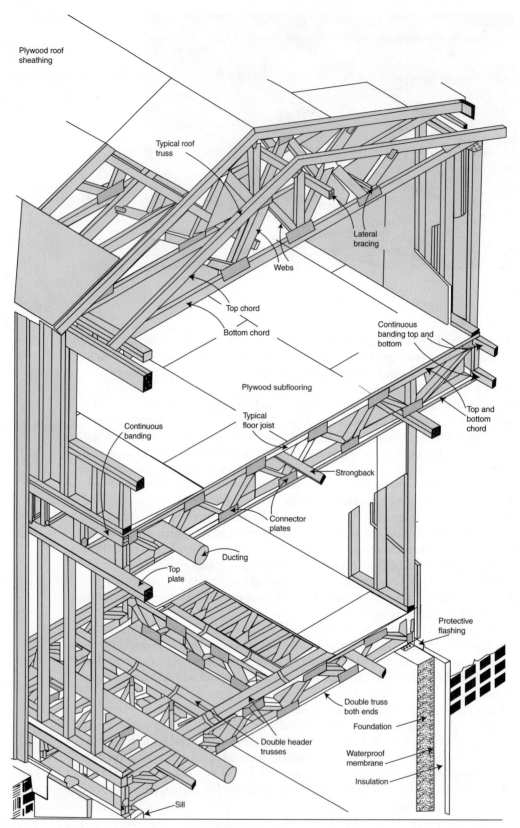

Plywood roof
sheathing

Typical roof
truss

Lateral
bracing

Webs

Top chord

Bottom chord

Continuous
banding top and
bottom

Plywood subflooring

Continuous
banding

Typical
floor joist

Top and
bottom
chord

Strongback

Connector
plates

Ducting

Top
plate

Protective
flashing

Double truss
both ends

Foundation

Double header
trusses

Waterproof
membrane

Insulation

Sill

FIGURE 7-20 Any connection plate (gusset) failure will lead to failure of truss. It is not good to be on top or under a truss when this happens.

becoming concerned that their building might be construed as unsafe; this creates angst in some politicians. To protect the lives of firefighters, marking the buildings is certainly a worthwhile endeavor. Research those communities that have been successful in passing the pertinent legislation and see if you can employ the same strategies in your community.

ENGINEERED WOOD LUMBER

The term engineered lumber covers a vast array of manufactured wood products. It is mainly used to describe structural components. These include beams, truss joint I-beams (TJIs), laminated strand lumber (LSLs), parallel strand lumber (PSLs), laminated veneer lumber, and plastic wood.

Engineered wood is manufactured by bonding together wood strands, veneers, lumber, or other forms of wood fiber to produce a larger and integrated composite unit that is stronger and stiffer than the sum of its parts.

LAMINATED VENEER LUMBER

Laminated veneer lumber uses multiple layers of thin wood joined by adhesives. The layers of wood are arranged in plies similar to plywood, however, they all face in the same direction rather than alternating plies perpendicularly. This product is often used when bending is needed as in the creation of an arch. This process would include microlam and **glulam beams**.

glulam beams ■ A type of structural timber composed of layers of glued laminated timber.

LAMINATED- AND PARALLEL-STRAND LUMBER

Laminated- and parallel-strand lumber differs in the application of the wood fibers. Laminated-strand lumber uses small chips or strands similar to OSB while parallel-strand lumber uses longer strands with more inherent grain defects in order to create a random pattern in the wood. LSLs and PSLs are often clad in a veneer to replicate a certain species of wood such as oak, cherry, or maple, further disguising their true composition.

The advent of these processes, and why they will continue to proliferate, is due to the decreasing amount of old growth forest in this country. The forest industry today uses a process called *forest farming*. This process controls the planting and harvesting of trees solely for the purpose of supporting the building industry. This process also allows for the total use of all parts of the tree. The tree will be sawn at the mill for parts, the remaining bark will be processed for mulch, and any remaining parts of the tree will be ground up and used in the manufacture of plastic wood, OSB, and other products.

Building Green

PLASTIC WOOD

Plastic wood is available in multiple forms, which are manufactured by several different processes. The forms include composite, extruded, and a combination of foam and plastic. These products are currently used outdoors, for example, in decks and furniture. I would not be surprised, however, to see them soon being accepted by the major code groups for use in structural framing.

Composite wood is made from a mixture of sawdust, recycled materials from used wooden pallets, recycled plastics from industrial waste, and recycled plastic bags. The extruded wood consists of recycled plastics from industrial waste and fiberglass.

The third form of plastic wood is a combination: a hard plastic shell with polyurethane filling. This product is used mainly for trim. Once painted it is almost impossible to detect that is not wood. The problem that the use of plastic wood creates for firefighters is the amount of energy produced by plastics as they are consumed by fire; all burning plastics produce 18,000 BTUs, so as the fire loading is increasing, the fire intensity is also increasing.

SCHOOL OF HARD KNOCKS

I have always had a healthy respect for trusses, whether they were metal or wood. I am especially aware that it is only when the decking and truss have been joined that the system is complete. My units were combating a fire in a dwelling that contained both trussed floors and roof. The fire occurred at about 9:00 A.M. on a weekday. Three rooms on the second floor were involved. I advised all units to use caution and gave the order that no truckmen were to go onto the roof surface. We had been alerted that the occupants had already escaped. The units had made the ascent up the stairs and were entering the hallway when the first partial collapse occurred in a room adjacent to the ones involved. I withdrew the units before the complete failure took place. After witnessing firsthand the devastation and the relative speed with which the failure occurred, I have come to the conclusion that if the occupants are out of the structure it is best to attack the fire from the exterior. I would do this for all commercial structures. This may smack against tradition, but our tradition is killing our own too often and for no valid reason.

FIRST RESPONSE

Operating Safely Around Trusses

There is no doubt that trusses will be found in all parts of the United States. You will find them in residential buildings as floors and roofs. You will find them in older commercial and public buildings as monster roof assemblies. You will find them in churches and other places of worship.

The good news is that for the past twenty years most codes have required metal trusses in most publicly occupied buildings, including schools, government offices, arenas, and most commercial occupancies. Check your local code. The metal trusses will appear as roof structural members, floor supports, and also, in an exoskeleton-framed building, as major supporting frames on the exterior of the building.

There are people preoccupied with a time-driven margin of safety, a margin that dictates how long firefighters can safely operate before truss failure. There is no margin of safety driven by time. Depending on the age, integrity, and configuration of the trusses, both metal and wood, the amount of flame and heat coupled with the load will dictate the failure.

Some tactical and strategical considerations are listed below:

- The only time it is justified to risk a firefighter's life is to perform an actual rescue or conduct a search where the victims have been seen, heard, or reliably reported.
- No commercial or residential building needs to be entered to successfully extinguish the fire within.
- If a search or rescue needs to be accomplished within a trussed building, then hoselines need to be advanced to control the fire and these lines should accompany the firefighters performing the rescue or search.
- For all unoccupied or vacant commercial structures with significant fire within, the first action to be taken on the scene is to establish the collapse zones. Indicate where they are with tape, rope, or some other means so that no fire suppression activities occur within the confines of the zone.
- A thorough size-up report from all sides of the building should be conducted prior to making tactical deployments.
- Additional safety personnel should be on-scene of a trussed building.
- Any time entry must be made, the entering firefighters must stay near the bearing walls and use the reach of the stream to combat any fire.

- Don't be lulled into a false sense of safety. If you see heavy timber trusses, assume that they will also fail.
- Conduct familiarization inspections for all commercial buildings and other target hazards in your area.
- Prepare pre-fire plans according to NFPA 1620: *Standard for Pre-Incident Planning*.
- Pursue legislation in your community requiring that all buildings containing trusses be marked with an icon that is visible at night.
- Make safety the first priority for all responders at these structures.

ON SCENE

Your company is assigned to perform a familiarization inspection of several buildings under construction. The floor and roof assemblies are constructed of lightweight truss construction. The exterior walls and roof deck assemblies are sheathed with OSB. The buildings, when completed, will be equipped with sprinklers in the attic.

1. What firefighting concerns can be associated with this construction?
2. What problems can be addressed by your department's standard operating guidelines?
3. What can your local building code enforcers do to notify all future responders that these conditions exist?

Summary

Trusses have become common in the building industry in all types of construction, so it is important to understand not only how they work, but how they fail. Whether they are metal or wood, when subjected to the intense heat of a fire trusses will give way, creating potentially dangerous collapse situations. If you understand the forces acting upon trusses, it can help you to be safe when you are fighting fires.

The key to any operation involving trusses is to respect them and to understand that they will fail. It's just a question of when—and where you are—in relation to the failure.

It is also important to identify buildings that use truss systems so that in the event of an emergency, caution can be taken. Some fire departments have been able to get buildings with trusses marked with an icon.

By understanding truss systems, you can ensure that those who work under your command can fight fires more safely.

Review Questions

1. Define the parts of metal and wooden trusses and compare the forces acting on each.
2. What could be a better method for attaching wooden truss components?
3. Why are wooden trusses better in compression than tension?
4. Discuss how metal trusses cause masonry walls and columns to fail.
5. Discuss the evolution of trusses in commercial structures.

Suggested Reading

Dunn, V. 1988. *Collapse of Burning Buildings*. Saddlebrook, NJ: PennWell, 1988.

Hibbeler, R. 1995. *Structural Analysis*. Upper Saddle River, NJ: Pearson Education.

NFPA 251. 2006. *Methods of Tests of Fire Endurance of Building Construction and Materials*. Quincy, MA: National Fire Protection Association.

NFPA 1620. 2010. *Standard for Pre-Incident Planning*. Quincy, MA: National Fire Protection Association.

Onouye, Barry S.; Kane, Kevin. 2001. *Statics and Strength of Materials in Architecture and Building Construction*, 2nd Ed. Upper Saddle River, NJ: Pearson Education.

Smith, Ronald C.; Honkala, Ted L. 1994. *Carpentry and Light Construction*, 5th Ed. Upper Saddle River, NJ: Pearson Education.

8

Size-up

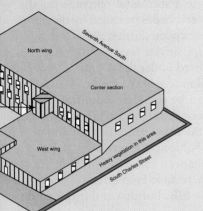

NFA formula, *p. 189*   size-up, *p. 186*   triangle of potential, *p. 190*

OBJECTIVES

After reading this chapter, you should be able to:

- Understand the concept of performing size-up.
- Make better decisions regarding entry into structures.
- Calculate what resources will be needed for a safe and effective attack.

Introduction

The first arriving unit or units at a scene provide confirmation of an incident. The brief initial radio report alerts all responding units, including the incident commander, of the scale of the incident. In many instance, it also confirms the address. The information contained in the brief initial report (BIR) helps determine a successful outcome for the incident. Some departments use 10 codes. Prescribed numeric codes preceded by the number 10 convey facts as defined by that department. These codes shorten radio messages normally given in narrative form. Unfortunately, the fire service has not standardized the use of 10 codes. Their use has therefore been disallowed under the conditions of the presidential directives governing the National Incident Management System (NIMS). For many departments with only one radio channel there can be so much radio traffic that many important messages are not received or they are not received in a timely fashion. For this reason an initial, brief, verbal narrative is now the usual procedure. It is important that this report be both specific and informative. Subjective terms such as heavy, light, or moderate are based on a perception of conditions seen and don't provide information about the condition of the structure involved. There have been attempts over the years to include building construction type as part of the BIR, but that is difficult to do accurately regarding today's building techniques.

 FIREHOUSE DISCUSSION

The fire at the Pang warehouse in Seattle, Washington occurred on January 5, 1995. This fire would cost the lives of four veteran Seattle Fire Department firefighters. The building was vacant at the time of the fire. This fire underscored many important and valid concerns for the fire service.

- Need to perform pre-fire plans for major buildings
- Need to make a complete and accurate size-up
- Need to make progress reports to command at regular intervals; if not forthcoming, the incident commander must request them
- Need for subsequent responding chief officers to conduct their own complete evaluation of the scene by physically traveling around the affected structures and then evaluating operations
- Safety officers need to versed in building construction and experienced in fire attack operations

The Pang building was constructed in 1909 as a single story 60' × 60' structure. Two 60' × 60' additions were later added on the west and north sides of the original building, forming an L shape (see Figure 8-1). In the 1920s, the city of Seattle undertook the daunting task of raising the ground around the structure 20'. The building had been located in low-lying swampland and the original building had become inundated with earth. The east wall was raised 5' in height to act as a retaining wall. The remainder of the structure became a windowless basement only accessible from the west and north additions. Another addition was added onto the original section, and this became the street level entrance. The building appeared to be single story due to the topography (see Figure 8-2). The south side appeared to be on an incline, but this view was obstructed by heavy foliage. The front of the building was disguised by a stucco paneled exterior. The west side of the building was the only vantage point from which the complexity of the building could be determined, including making evident the existence of multiple stories.

The original building had masonry exterior walls. Two of these walls became interior party walls when the additions were added. The roof of the original building was constructed using 12" × 12" posts. The second story center section addition utilized 2" decking resting on 3" ×14" wood joists supported by 8" × 14" wood beams. These beams were set in cast iron caps placed on top of the original posts. The original roof had rested on a narrow ledge set in the masonry walls along the original north wall. The new floor of the center section addition was higher than the original wall so a knee wall or *pony wall* was constructed from 2" × 4" sawn studs; this wall rested on the original masonry ledge (see Figure 8-3). A pony wall is a wall that is less than 8' in height. It is often used as a partition within a room or space to differentiate two spaces. A concrete floor approximately 1" deep had been added to the wood decking. The roof was comprised of tar and gravel and was set on the roof joists. A wooden ceiling had been installed on the interior.

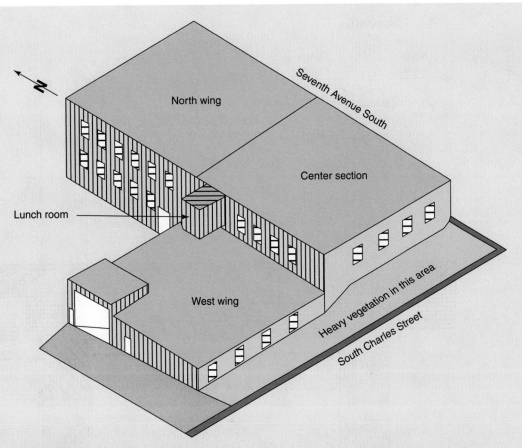

FIGURE 8-1 View of building showing western view.

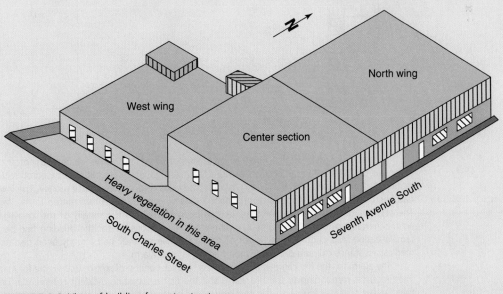

FIGURE 8-2 View of building from street entrance.

(continued)

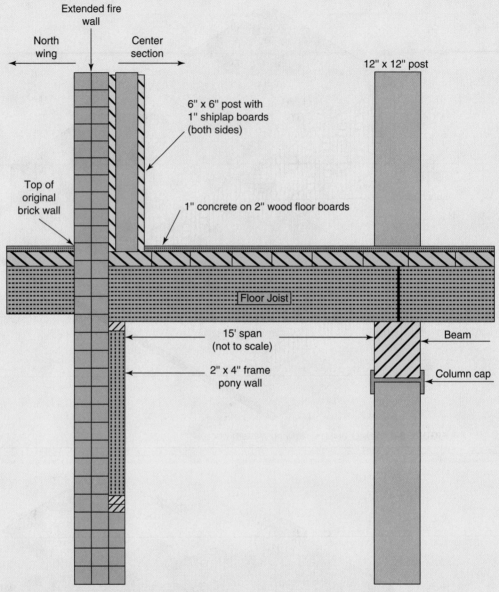

FIGURE 8-3 Elevation view of pony wall assembly.

The building had been the subject of an investigation by the Seattle fire department arson section as well as the Federal Bureau of Investigation. The investigations were initiated based on reliable information that the owner's son was going to burn the building for the insurance money. The investigation had been called off and no further information regarding it had been transmitted to the department. The first responding incident commander had been briefed on the details of the investigation and actions pertaining to it. The engine company normally responsible for the building had performed a familiarization inspection. On the night of the fire, however, they had already been dispatched on a medical assistance call and did not respond until the third alarm.

The alarm for the fire came in at 1902 hours. A thick column of smoke was visible for many blocks. The first units responding to the fire alarm and the first responding incident commander determined that the fire that was visible on the east side of the center section addition and it was actually on the exterior of the building (see Figure 8-4). The first-in company officer and the incident commander jointly decided that the engine company enter the structure and fight the fire from within, east to west, to keep it from entering the structure. The Incident Command System was implemented and

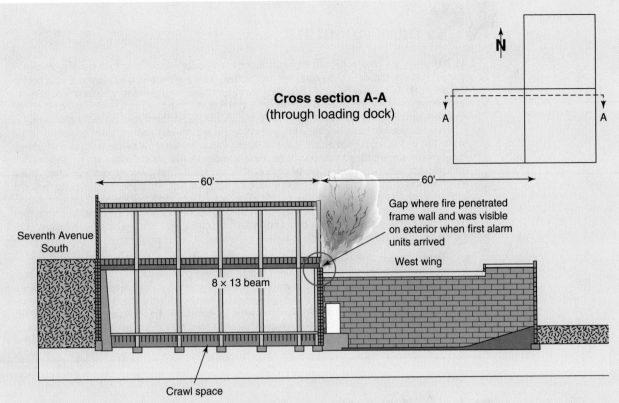

Cross section A-A
(through loading dock)

60'

60'

Seventh Avenue
South

Gap where fire penetrated
frame wall and was visible
on exterior when first alarm
units arrived

West wing

8 × 13 beam

Crawl space

FIGURE 8-4 Cross section of area where fire was visible to units arriving on-scene.

assignments were given. A chief officer was charged with responsibility for the accountability system. The subsequently responding command officers and the designated department safety officer reported to the command post. All agreed and approved the tactics and strategies that had been developed by the initial IC. At the time of the disaster, the incident commander and the safety officer were about to commence a size-up circuit of the building because this had not yet been accomplished. The units who had entered the center section addition interior building encountered intense heat and smoke, but very little fire. The fire, which was visible at the floor level, was quickly knocked down. The initial ladder/ truck company went to the roof of the center section addition and witnessed fire lapping over the roof on the west side. They then began to initiate a trench cut, an approximately 3' to 4' wide opening in the roof that runs the width of the structure. Although this operation is very time consuming, it should be attempted as a defensive maneuver.

The engine company assigned to the rear (Division C), in the rear of the building, discovered a heavily involved room measuring 30' wide × 60' deep and 20'. They had been ordered to hold their position and assumed that firefighters would be coming towards them making an aggressive interior attack. Neither the truck company on the roof of the center section addition nor the engine company reported their findings to command, nor did the IC request reports on conditions at their locations. The command officers began to believe that there might possibly be a fire in the cockloft due to the increasing thermal column rising above the structure, the lack of heat in the center section addition, and the fact that the fire was not yet extinguished.

At 1932 hours the incident commander requested a third alarm. Just before the disaster there were three lines being advanced on the center section addition interior. Approximately twenty-nine minutes after the first size-up report given by the initial incident commander, the pony wall failed and eleven firefighters were cast into the now raging inferno. Seven firefighters made it to safety but all were burned. Four firefighters remained trapped in the basement of the center section addition.

Fourth and fifth alarms were requested. The accountability system worked very well and within minutes all firefighters were accounted for as either still trapped or on the exterior of the building. Heroic efforts were made in the attempt to rescue those trapped but to no avail. The rescue attempts and the subsequent body recovery operations were managed extremely well using the Incident Command System. All four firefighters succumbed to asphyxiation. All four were wearing manually actuated PASS devices. Two were turned off.

All firefighters are constantly torn between making aggressive interior attacks and being prudently safety conscious. This dichotomy continues to this day. There are those in the fire service who believe that the faster water is applied, the faster the fire will be extinguished giving problems no time to manifest themselves. There are others who have come to believe that no building is worth risking firefighter lives; these individuals would prefer all building fires should be extinguished using outside streams.

Risk management involves the initial size-up by the first arriving unit and another size-up circuit by the first arriving command officer followed by correlation between the two reports; also taken into consideration are reports from other units on the roof and basement areas. Subsequent progress reports from all supervisory personnel should be requested by command if not given without prompting. These reports should indicate whether the mission could be accomplished with present staffing and tactics, if additional units are needed, or if a change of tactics is called for. Additional arriving command officers should also perform a size-up circuit around the building. Once returned to the command post, they should make note of the amount of time units have been operating and make recommendations as to the validity of the current strategies and tactics.

- Discuss fighting fires in vacant or unoccupied buildings as pertains to risk management.
- Discuss how frequently progress reports should be given or requested.
- What steps should an incident commander take when conditions change or progress is not being made?
- Discuss the role of the safety officer and what qualifications are necessary to be effective. For example: should a safety officer necessarily be an officer; how much construction experience or education is necessary; how much political experience is essential.

Phases of Size-Up

size-up ■ Estimation of problem faced or ongoing evaluation of operation; includes brief radio reports between command and units.

According to the National Fire Academy there are three phases of **size-up**. Phase I is pre-incident information. Phase II is initial size-up, which is what you see and hear from the time of dispatch until you complete size-up upon arrival at the incident. Phase III is ongoing size-up, the gathering of information throughout the incident. Ideally, Phase I should include information collected long before the incident takes place. The ability to fight fires safely depends in large part on what experience and knowledge is possessed before the incident. This is the information that allows you to make informed and intelligent decisions upon your arrival at the scene.

PHASE I: PRE-INCIDENT INFORMATION

Most firefighters think Phase I involves knowing where they are going and what they will need. However, as you have read in the previous chapters, a solid basic knowledge of building construction is paramount in making reliable and accurate size-up. The value of studying building construction and its relationship to fire behavior needs to be emphasized within the fire service, because—obviously—buildings are the principal environments in which the fire service operates. Understanding building construction—and the ability to size up a building as part of a firefighting plan—can help prevent firefighter fatalities.

The building industry is constructing buildings quickly and inexpensively, and new construction materials make it harder to determine if the construction is a Type III or a Type V with an elaborate veneer. Understanding how a building is constructed certainly helps to predict how and when it will begin to come apart during an emergency. For example, in a Type V with a veneer the connection between the masonry veneer and the wood framing will disintegrate. This will contribute to early failure of the masonry, especially the chimney, and will put firefighters at great risk.

Without this knowledge a firefighting plan is meaningless. Most pre-plans currently address only resources and not potential failures based upon the building's condition. Pre-plans need to address the collapse potential based on construction features as well as the resulting potential for building failure. For example, an explosion in a Type I (rated fire resistive) does not affect the building in the same way that fire does. This is discussed in Chapter 9.

PHASE II: ARRIVAL

Size-up, or incident analysis, is a key part of Phase II. It is an evaluation process that considers all factors that can have a positive or negative impact on the situation. Size-up is a mental process involving the quick weighing of the factors found at an incident against the available resources. During size-up the condition of the structure is also assessed. This sounds complicated but it is only an assessment of what you see. As soon as practicable the facts need to be evaluated relative to construction type and potential for failures.

A common mnemonic (WALLACE WAS HOT), which is used to assist the incident commander with size-up, is shown below. You can also use the acronym COAL WAS WEALTH.

Water	Weather
Area	Auxiliary/Appliances
Life	Special Matters
Location, Extent	
Apparatus	Height
Construction/Collapse	Occupancy
Exposures	Time

Size-up should consider more than what you see through the windshield upon arrival. The initial size-up must be done in a calm, objective manner, ensuring that the officer first identifies the problem(s) before applying the solution. The brief initial report (no longer than 3 to 4 seconds) should paint a vivid word picture of the conditions. This gives the other responding units a clear understanding of the extent of the incident and gets them up to speed. If action precedes size-up the safety of personnel is jeopardized and resources may not be utilized to their maximum effectiveness.

The structure itself must be described in as few words as possible while still being accurate. If you provide the following information in describing the structure upon your arrival you will have satisfied many of the facts needed for accurate size-up reports.

Height
Size (given as width × length)
Use/configuration
Scope of problem observed
Special issues

Give a group of fire service personnel this information for any address in their area and they will tell you what is burning and how much equipment it will take. The first three items are self-explanatory. The scope of the problem needs to be defined by the incident commander before the response. For example, many firefighters might describe a volume of fire or smoke as heavy. This is a subjective term and will differ between individuals based on experience, knowledge, and state of mind at the time of the report.

As for special issues, too often we try to guess the type of construction. But this should be definitively identified right up front. We've already discussed how many types of construction can have exactly the same type of outer covering, which can include masonry, stone, stucco, DryVit (foam panels installed with adhesive, screws, or nails and covered with an epoxy to mimic stucco), or some form of siding installed over any of the above. Confirmation has to be made by a unit reporting to the basement area to ascertain and confirm the building's construction type. The apparent structural condition must also be evaluated at this point. Any smoke appearing to pass through masonry, any flame volume sufficient to impact structural components, or any unusual appearance contrary to the normal structural configuration of the type are evidence that firefighters could be ambushed by structural failure. Reports of civilians trapped or missing, including their location, must also be confirmed. For example, a report of persons trapped on the third floor while the fire is reported to be on the second floor is certainly a different situation than if the people were trapped on the first floor due to smoke conditions while the fire was on the second floor.

Once the initial size-up is completed, the incident commander should call for sufficient resources to handle current and potential problems. If there is any doubt as to the structure's integrity, the IC must take a defensive stance immediately. Two more acronyms—STOP and WHOA—can used when making this evaluation.

Both of these acronyms emphasize the importance of evaluation: first slow down and look then decide if there are enough resources, of both personnel and water, on the scene or coming. Ask yourself if the building is safe to operate in. This evaluation must be accomplished in a compressed time period.

Stop	What do I have
Think	How long has it been going on
Observe	Outcome if I go defensive
Predict	Affect on the structural integrity now

The fire service has always operated under the assumption that if you can get into service fast enough, you can stop the incident while it is still at a manageable level. The ability to ascertain how much water and how many lines are needed to accomplish the task is a critical part of the decision process. This is described next.

NATIONAL FIRE ACADEMY FLOW FORMULA AND IOWA RATE OF FLOW FORMULA FOR FIRE CONTROL

There are two methods recognized by the United States fire service for determining needed water flow. These are the Iowa Rate of Flow Formula and the National Fire Academy's Fire Flow Formula.

Keith Royer and Floyd W. Nelson conducted research for Iowa State University. Their research objective was to identify what happens in uncontrolled fires in structures and why fire behaves the way it does in these environments. By 1952 they had gained enough practical knowledge to begin researching the application of water in these scenarios. From 1955 to 1959 they conducted hundreds of experiments to determine the amount of water needed to control fires in all sizes of buildings. The key is that this formula is designed to control a fire, not extinguish it.

The formula used to calculate required water flow is length × width × height divided by 100 assuming an estimated time of delivery of thirty seconds. The calculation determines the amount of water necessary to knock down the fire in the largest single, fully involved, open area. It does not allow for water usage needed for exposures or in other rooms. The formula does not allow for the conversion of water to steam. The premise was that steam conversion is only germane within closed compartments, which was the basis for Lloyd Layman's contribution to the fire service regarding fog streams.

Royer and Nelson's formula is thought to be a good planning tool but is not necessarily useful in tactical applications. The basis for the formula is knock down, not extinguishment, and assumes that this had to occur within the first ten minutes after the fire started. Their formula is all or nothing: If resources or equipment are not available within this time frame, it is their belief that the building will be lost.

The National Fire Academy (NFA) has devised a method for the fast computation of water needs. The NFA formula can be used as a planning tool and a tactical tool. There are limiting factors, however:

- The formula is designed for offensive, interior operations involving direct attack.
- The formula is inaccurate at flow rates greater than 1,000 gpm or for involvement greater than 50 percent of the structure.
- The formula is based on area, not volume. Ceilings more than 10′ high can have a pronounced effect, resulting in underestimation.
- The formula has not been adjusted to account for today's higher heat outputs from fuel loads, an increase from 8,000 BTUs in 1980 to 180,000 BTUs today.

NFA formula is (width × length)/3 plus exposures (25 percent of resultant for each exposure) minus the percentage of involvement (the formula is predicated on 100 percent involvement. So a two-story building 100′ × 50′ with 25 percent involvement on the first floor would require approximately 500 gallons of water:

$$(100 \times 50)/3 = 1666 + (1666 \times .25) = 2082 \times .25 = 520 \text{ gallons}$$

NFA formula ■ Developed by the National Fire Academy to calculate quick estimates of resources at the scene.

With practice, this becomes simple if you round off numbers. Many of our pumping apparatuses have 1,000 or 1,500 gpm pumping ability. Due to staffing, however, most engine companies can support only one line. If you're using a 1½″ hoseline you can provide 100 gpm; with a 1¾″ line you can provide 200 gpm. So for every 100 to 200 gallons needed you must have one engine company. Therefore, the size of the burning building and the extent of involvement are critical in the process of size-up. For this reason, the developers of the NFA course, *Preparing for Incident Command,* use the NFA method when teaching students how to calculate the amount of water flow necessary to conduct an aggressive direct interior attack. To learn this method, various scenarios are presented to Academy students. The students are told to control the fire using the methods their department would use, while emphasizing number, placement, and flow rate of handlines. The scenarios are limited to no more than 50 percent involvement of a building by fire.

PHASE III: ONGOING

Reappraisal of the incident scene should be continual. Reassessment, done in Phase III, will keep the incident commander abreast of changing conditions and aware of where resources may have to be redeployed. Most incidents are residential and require 100 to 300 gpm of water deliverance for about five to ten minutes. The first arriving units and the IC must assess the fire's size and its impact on the building.

Size-up does not stop once firefighters implement the action plan. Size-up needs to be ongoing throughout the entire incident. The initial information available to the incident commander is often incomplete or wrong. Reports must be given to the IC at regular intervals from the companies operating inside the building. The failure to communicate significant information up and down the incident management structure is one of the most common breakdowns leading to firefighter injuries and fatalities.

Who should make these reports? The IC must prompt this information gathering, contacting units in all parts of the structure and periodically asking, "What is your location and what are the conditions?" This information will provide an accurate and ongoing analysis of the problem's growth or containment. It will also allow the IC to withdraw the units faster if the incident escalates to the point of potential failure. If the IC waits for information to come in, the incident may escalate to a critical point before pertinent information reaches the IC. Think of command as playing chess rather than checkers: In checkers the players react to moves; in chess, strategy is planned several moves ahead.

Post-incident analysis regarding fatalities often reveals that individuals on the scene made critical observations but neglected to effectively communicate the information to the IC, therefore the incident commander was not aware of potential disasters. On January 5, 1995, four firefighters were killed in Seattle, Washington. They died when a floor collapsed without warning. The building had access from different levels on different sides of the building, resulting in confusion as to the level at which the units were operating as well as their position in relation to the location of the fire. Because of this some units thought they were safe when they were actually directly above the fire.

Developing Size-Up Skills

The ability to give a clear, concise size-up report is one of the single most important skills a company officer must master. Unfortunately, most company officers do not get to practice this skill on the fireground very often. The solution is to conduct a size-up exercise. The skills that must be reinforced include recognition of various construction types and the hazards associated with them, the use of the order model (a mathematical model whereby sufficient resources are brought to bear to solve a problem in a timely fashion) and communications procedures, and the use of staging procedures and command functions. Practice until you are comfortable with these.

Then practice ongoing size-up. This drill does not have to be complicated. You can use PowerPoint slides showing different scenarios. There are programs available that allow the use of photos showing local buildings. The trainer applies animated smoke and flame. The trainer can then evaluate and critique your size-up.

Building construction should be a major component in every training regimen. This is the work site for anyone involved in the fire service. An officer cannot be expected to give accurate size-ups if they have no idea what they are observing. An incident commander cannot be held accountable if they have no experience at the building at which they are commanding troops, or if they have not had the prerequisite training to recognize dangers inherent to that structure.

POST-INCIDENT ANALYSIS OF SIZE-UPS

All working fires or other significant incidents must be critiqued. The critique must contain nonattribution rules. A neutral chief officer should chair the proceedings to allow more freedom of communication. The critique should not become a spin or blame exercise. Critiquing allows members to learn from positive and negative actions taken at each incident, thereby allowing firefighters to identify early indications of problems with tactics, procedures, or equipment.

TRIANGLE OF POTENTIAL

Gordon Graham is a risk management expert and former California Highway Patrol officer. He and several others have developed a close call reporting system for firefighters who are seriously injured or almost killed while operating at an incident. Graham came up with a statistical matrix to predict close calls. Here's how it works: Envision a pyramid-shaped triangle. The base of the pyramid represents 300 occurrences of unsafe acts, poor size-ups, or other divergences from good working practices. That then leads to the middle section of the pyramid with thirty near-misses or close calls. If you don't correct the mistakes then it leads to the pyramid's peak of one major hospitalization or fatality among your firefighters. This is called the **triangle of potential**.

Reporting close calls and addressing their causes can help save firefighters from injury and death.

triangle of potential ■ Statistical matrix calling for increasing chances for injury or death if behaviors are not corrected.

SCHOOL OF HARD KNOCKS

The District of Columbia is divided into four sections: northwest, northeast, southwest, and southeast. All numbered streets are repeated throughout the four sections. For example, there are four 500 blocks of 12th Street, each designated as either northwest (NW), northeast (NE), southwest (SW), or southeast (SE). The lettered streets follow the same pattern. The residential building stock is primarily Type V and Type III. Most of it is configured into rows.

Throughout my tenure as a company officer, I always included the size of the building, the height of the building, and a clearly stated mnemonic for the numbers in the address when I gave the Brief Initial Report or BIR. This prevented any miscommunication with the communications section or other responding units.

When I became a chief officer, I instructed my units as to how to communicate with me in terms that I would understand. I also made sure that they understood our relationship on the scene; it was imperative for them to know that if I had any concerns as to the safety of the structure, I would pull them out immediately. I was prepared to err on the side of safety. This front-loading of information greatly assisted in having safe and effective firegrounds.

We were operating at a three-story end unit of Type III construction. The fire was involved on all three floors. I had been parked parallel to the structure. At this point the units were making good progress. I had units in adjoining exposures pulling ceilings and truck companies on the roof performing vertical ventilation. Suddenly a ball of smoke pushed out from an exposure and then went back in. I immediately withdrew all units. To this day I can only state that the ball of smoke was unusual and that's what made me call for withdrawal. As the last firefighters were leaving via ground ladders, a fireball blew out from the upper floors of that building, and I knew I had made the right decision.

FIRST RESPONSE

Nursing Homes/Extended Care Facilities

Nursing homes and extended care facilities share one important trait. They both contain occupants who will need rescuing or, at the very least, assistance in exiting the building. Any event in one of these occupancies should be considered a mass casualty event. The fire may be small, but it will be the smoke that will do most of the damage. These occupancies are most likely to be Type I and Type II structures; however, older facilities may be found in any of the building types. The facilities may contain sprinkler systems and standpipes but that will depend on the strength of your local building codes. They will have oxygen stored on the premises and oxygen in most of the rooms. They may also contain other chemicals on site, some of which may be flammable. The staff may or may not be trained to assist in the evacuation of patients. Very few rescues can be made from ladders due to the infirmity of the occupants. Many of the occupants will not be able to respond to verbal commands or directions. Weather will play a large role in your rescue efforts. If it's too hot or too cold or if it's raining, the rescued individual(s) might not survive. These occupancies can certainly be considered as a target hazard.

Some considerations for strategies and tactics are listed below:

- Have a pre-plan for these buildings and ensure all potential responding units have a copy.
- Institute a mass casualty plan.
- Identify resources, such as buses with handicapped access, in which to place the victims after you have removed them from the building.
- Institute the incident command system and understand the role of a branch director.
- Ventilate simultaneously with other operations.
- Protect all stairwells against fire spread if the building is not sprinklered.
- Institute mutual aid plans or multiple alarm plans as part of your pre-plan.
- Know the patient capabilities of the closest medical facilities.

A fire has occurred in a two-story 19′ × 33′ row house. There appears to be smoke coming from the front door. The occupants of the building indicate that everyone is out of the structure. There have been no reports from the rear of the structure. The time is 2330 hours. Upon entering the building the engine companies are advancing 1½″ attack lines but have yet to find the fire.

1. As the arriving incident commander, what concerns do you have that must be addressed immediately?
2. What does the information given by the occupants have to do with firefighting activities?
3. What does the time of day have to do with firefighter safety?

Summary

Size-up is key to firefighting safety. There are three phases of fire incident size-up and how the information is assessed and conveyed helps determine the success of the entire operation. The brief initial report in Phase I should be no more than four seconds in length and should provide information that is specific and informative.

Phase II size-ups should be done calmly and should paint a word picture of the incident. They should describe not only the size, shape, and scope of the problem, but should address any special concerns that might affect firefighter safety. Clarity and conciseness are crucial in the size-up.

The size-up should be ongoing, continuing into Phase III. Once firefighters are deployed, the IC should be in touch with them and should know their locations as well as the conditions in their area. Practice exercises can be used to gain more experience doing size-ups.

The triangle of potential shows that mistakes in size-up lead to injury and death of firefighters. By assessing conditions at the incident and keeping on top of changing conditions, incident commanders can deploy their resources in the safest way possible. Today's ICs need to possess a thorough knowledge of construction techniques and the vulnerabilities of the materials used. During an incident, they must stay in touch with all fire crews under their command, asking where they are and what the conditions around them are. In this way, they can take immediate action should problems occur.

Review Questions

1. Discuss which method of size-up your department uses.
2. Why would the word *heavy* be confusing in a BIR?
3. Discuss utilizing the NFA formula and provide examples of buildings that would overwhelm your department at only 25 percent involvement.
4. Why is information important when developing strategies and tactics?
5. Discuss pre-plans and how they would assist in performing size-up.

Suggested Reading

Brunacini, A. 2001. *Fire Command*. Phoenix, AZ: Heritage Publishers.

Dunn, V. 1988. *Collapse of Burning Buildings*. Saddlebrook, NJ: PennWell.

Flin, R. 1998. *Sitting in the Hot Seat*. New York: John Wiley and Sons, Inc.

Iowa State University Bulletin #18, 1959. *Water for Firefighting—Rate of Flow Formula*.

Nalder, E. and Wilson, D. The Pang Fire: What Went Wrong—A Disaster Marked by Bad Preparation, Poor Communication, and Many Other Mistakes. *Seattle Times Newspaper*, June 11, 1995.

Nelson, F. 1991. *Qualitative Fire Behavior*, Asland, MA: International Society of Fire Service Instructors.

Routley, J. G. 1995. *Four Firefighters Die in Seattle Warehouse Fire*-Report #077. Seattle, WA: United States Fire Administration.

Terpak, M. 2002. *Fireground Size-Up*. Saddlebrook, New Jersey: PennWell Books, 2002.

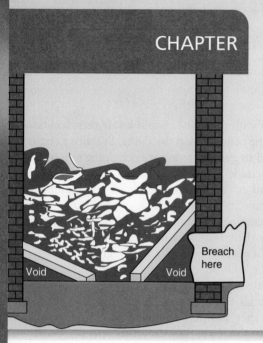

Void   Void

Breach here

Collapse

collapse zones, *p. 206*

curtain wall collapse, *p. 202*

deflagration, *p. 199*

detonation, *p. 199*

inward/outward collapse, *p. 202*

lean-to collapse, *p. 200*

90-degree collapse, *p. 202*

pancake collapse, *p. 200*

partial collapse, *p. 205*

voids, *p. 200*

V-shaped collapse, *p. 200*

OBJECTIVES

After reading this chapter, you should be able to:

- Understand the mechanisms of collapse.
- Predict potential collapse situations.
- Understand the various types of collapse.
- Recognize the many aspects of operations at collapses.

Resource Central

For additional review and practice tests, visit **www.bradybooks.com** and click on Resource Central to access book-specific resources for this text! To access Resource Central, follow directions on the Student Access Card provided with this text. If there is no card, go to **www.bradybooks.com** and follow the Resource Central link to Buy Access from there.

Introduction

Collapse is probably the most dreaded word on the fireground. Along with flashover it is one of the most dangerous conditions that firefighters face during their time in the fire service. Collapse occurs when external forces cause a portion of a structure, a system within a structure, or a total structure to fail. This chapter looks at the various types of construction and construction systems and examines why and how they fail. It examines size-up cues that assist in predicting when a failure may occur. It also looks at the mechanisms of collapse and their impact on building materials as well as what actions can be taken at the scene of a collapse. Finally it looks at the relationship between trained collapse specialists and local fire departments requiring their expertise.

 FIREHOUSE DISCUSSION

The Pittsburgh, Pennsylvania Fire Department has a long and storied history. On March 13, 2004 the members of the department fought a fire that cost them two dead and twenty-nine severely injured firefighters.

The fire was in a church constructed in 1875 and renovated in the 1930s and again in 1994. The main building measured 120′ × 70′ and was over 50′ to the roof line, bottom of the roof rafters, and exterior wall intersection. The exterior walls were comprised of several wythes of brick masonry covered with a stone façade that had been added as part of the 1930s renovation. The pitched roof assembly consisted of heavy timber roof trusses covered with sawn planking and asphalt shingles. A 115′ high bell tower with four spires had also been added during the 1930s renovation; it was attached to the main building using steel I-beams (see Figure 9-1). The tower constructed of exterior masonry walls covered with a stone veneer, provided support against the downward forces of the roof in the same way, as do buttresses.

The 1994 renovation included the addition of a three-story structure to the northeast corner of the church. This annex measured 60′ × 45′ and contained meetings rooms, a kitchen, bathrooms, offices, and an elevator.

The parishioners noticed smoke coming from an electrical outlet. They called 911 and evacuated the building. There were no occupants inside the building when the fire department arrived.

The initial engine company arrived at 0850 hours, only five minutes after receipt of the alarm, and advanced a 1¾″ attack line upon entry into the basement area. The firefighters were met by high heat and dark dense smoke. The first arriving ladder/truck company and subsequently responding ladder companies were ordered to operate on the first floor, opening it up for ventilation. The companies could not use their power saws due to insufficient oxygen and were performing the work using axes.

A crew entering the basement to assist with a handline broke a window at ground level for ventilation and noticed that the smoke emanating from the broken window was a dark yellow color and just didn't look right.

A crew responding on the third alarm was ordered to perform horizontal ventilation operations in the main church building. They were unable to accomplish this task because the large stained glass windows were covered with Plexiglas panels as a deterrent against vandalism.

At 0919 hours the incident commander evacuated all units from the building for an accountability check. After the check had been accomplished, the units returned to operations in the basement and first floor of the church.

At 0928 hours a backdraft explosion injured six firefighters. A backdraft occurs when a lack of oxygen creates an overabundance of carbon monoxide. The gases and smoke remain at high temperatures. Any infusion of oxygen, through a door being opened or window being broken, can cause sudden reignition as a result of the high heat and overpressurization. The result is usually catastrophic for anyone caught in the path of the gases seeking oxygen. The injured firefighters were taken to the hospital. Some were burned; others had been blown from their positions by the force of the blast.

At 0948 hours the decision was made to switch to a defensive operation utilizing master streams and large handlines. This was almost one hour after the units arrived on the scene. By 1048 hours the roof had burned off, exposing the remnants of the rafter trusses. The exterior streams were ordered shut down and the IC met with his staff to discuss overhaul. *Overhaul* is the process of removing debris and portions of the burned area to determine if the fire has been extinguished.

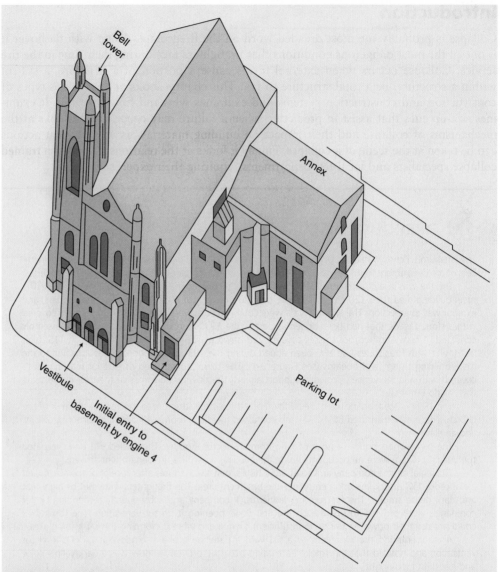

FIGURE 9-1 Diagram of the Pittsburgh Church showing addition and bell tower.

The two victims entered the main church building, or what was left of it, to check for any remaining pockets of fire. At 1213 hours the bell tower collapsed, trapping the two victims under tons of stone, brick, wooden rafters, and other debris. Units operating on the second floor of the annex were also trapped by falling debris. The IC, along with many of his staff, was struck by falling debris and became incapacitated. Their command post had been set up within the collapse zone, normally considered to be within one and a half to two times the height of the structure.

An accountability check was conducted. The identities of the missing firefighters were determined as was the status of the twenty-seven firefighters who had been injured in the collapse.

The two victims were killed by traumatic asphyxiation and blunt force traumas to their head and neck areas.

The church had been insured and was rebuilt in 2006. It included a room dedicated to the fallen firefighters.

Mechanisms of Collapse

A myriad of situations can result in a collapse: They can occur due to human error, such as poor workmanship, faulty design, or overloading, or from an overt attack, such as a bombing. They can be caused by weather-related events, such as tornadoes, earthquakes, floods, or hurricanes. Time and neglect can also be factors. A structure vacant and untended for a protracted period of time will eventually fail.

HUMAN-CAUSED MECHANISMS

Residential construction is exposed to far less scrutiny than commercial construction. The International Residential Code (previously the CABO One- and Two-Family Dwelling Code) was derived from the BOCA National Building Code discussed in Chapter 2 and is the prevailing code. The codes calls for plans to be reviewed by the appropriate building official relative to their compliance with the code, however, the work performed in the field is inspected and approved by a building inspector working under the authority of the state building official for the state. The following conditions are often missed (or are not required) during field inspections:

- Incorrectly sized nails or screws
- Inadequate nail patterns
- Inadequate support for main members
- Use of hollow masonry versus solid masonry in point loading of beams and girders
- Building not square
- Inadequate support for masonry fireplaces and chimneys
- Inadequate connectors for masonry veneers
- No testing of mortar mixtures
- No slump test required for concrete
- Removal of key components of support

These faulty workmanship issues will still meet the minimum requirements necessary to prevent collapse under everyday conditions, however, they will quickly add to collapse potentials when the structure is exposed to more extreme stimuli. For example, inadequate nail sizes and nail pattern spacing will cause a connection to fail much faster (see Table 9-1).

TABLE 9-1	Fastener Connectors for Structural Members	

STRUCTURAL MEMBER	NUMBER AND SIZE OF CONNECTOR	CONNECTOR SPACING
Floor joist to sill or girder	(3) 8d	Alternate every 14"–16"
1 × 6 subfloor to joist (pre-1960)	(2) 8d	Alternate every 14"–16"
2" subfloor to joist or girder	(2) 16d	
Sole plate to joist	16d	16" o.c.
Top plate to stud/end nail	(2) 16d	
Stud to sole plate/toe nail	(3) 8d or (2) 16d	
Double studs/face nail	16d	24" o.c.
Double top plates/face nail	16d	24" o.c.
Made-up header	16d	Alternate top and bottom 16"
Built-up corner studs	16d	24" o.c.
Built-up girder and beams	16d	32" o.c. at top and bottom and staggered 20d at ends and at each splice
2" planks	(2) 16d	At each bearing
Roof rafters to ridge/valley	(3) 16d or hip/end or toe nail	
Rafter ties to rafters/face nail	(3) 8d	

FASTENER SCHEDULE FOR STRUCTURAL WALL, FLOOR, AND ROOF ASSEMBLIES		

DIMENSIONS	NUMBER AND SIZE OF CONNECTOR (NAILS OR SCREWS)	CONNECTOR SPACING
Plywood subfloor, roof, and wall sheathing to framing		
4 × 8 × 5⁄16″ – 1⁄2″	6d	6" edge/12" intermediate supports
4 × 8 × 9⁄32″ – 3⁄4″	8d	6" edge/12" intermediate supports
4 × 8 × 7⁄8″ – 1″	8d	6" edge/12" intermediate supports
4 × 8 x× 11⁄8″ – 11⁄4″	10d	6" edge/12" intermediate supports
Wall sheathing other than plywood		
4 × D8 × 1⁄2″ fiberboard	11⁄2" galvanized	3" edge/6" intermediate supports
4 × 8 × 1⁄2″ gypsum	11⁄2" galvanized	3" edge/8" intermediate supports
4 × 8 × 5⁄16″ – 1⁄2″ OSB wall or roof	6d	6" edge/12" intermediate supports
4 × 8 × 5⁄8″ – 3⁄4″	8d	6" edge/12" intermediate supports

If the mortar mix contains too much sand, the masonry will not adhere as intended. On the fire scene, there are clues to this deficiency. If the masonry units in the debris field are clean, devoid of mortar, then the mix contained too much sand. The masonry should fail before the joint. The use of hollow block is fairly common in residential construction. The mason places mortar into the webs of the block as the wall rises because this allows for less cutting of block and is more economical, however, code requires solid blocks under point loads because the hollow block will fracture faster than solid block.

Most residential builders are conscientious, however, some builders will add more water to the concrete mix to enable the concrete to flow more freely through the forms. Less labor is then needed to push the concrete.

If you read a builder's contract for residential projects, you will usually see the phrase "good construction practices" throughout; these practices dictate how a process will occur.

The poor workmanship can contribute to failures when the end user, the occupant, overloads the system. For example, failure can occur when homeowners install waterbeds or hot tubs or allow too many people on an exterior deck. This is why architectural plans for these types of features seem to be overbuilt. The architect allows for the possibility of faulty workmanship.

A critical error in both residential and commercial construction occurs when key support members are removed before the entire system is stabilized. This often happens during remodeling or renovating. In the early 1970s the Vendome Hotel in Boston, Massachusetts, was being remodeled when a major fire broke out. Nine firefighters lost their lives when an entire section collapsed because a bearing wall had been removed in the basement.

Bombings or explosions can cause overpressurization either inside or outside of a structure, depending on the placement of the device. Overpressurization creates a **detonation**. The resulting shock wave, traveling faster than the speed of sound, causes connections to give way and the structure to fail. A backdraft can be considered a detonation. The 1995 bombing of the Murrah Building in Oklahoma City, Oklahoma, is an example of a detonation. **Deflagration** does not result in failure because the overpressurization travels slower than the speed of sound. A flashover is an example of deflagration.

Fires can cause failures. The structure and its systems are adequately supported against gravity until fire begins to attack. The thermal effects upon a building's structural systems can cause radically shifting loads. A beam or column can withstand the loads imposed in everyday use, but if another member can no longer support its share of the burden and fails, the subsequent loss of support will shift increased loads upon the surrounding members. Once failures begin the environment becomes a violent one. As columns, walls, and beams collapse they contact other members. The force of the contact can take the form of point loads, impact loads, or lateral shock loads. The result can be localized failures or complete failure of the systems within the structure, the latter causing the total structure to fail. This is particularly true in truss floors and roofs. The initial failure of one member of the truss creates a domino effect resulting in subsequent failures in other systems.

Firefighting operations can also cause failures within structures. The force of an attack line can dislodge mortar, master streams can knock down curtain walls and framed walls, and water collecting on roofs and floors can easily overtax the supporting members and assemblies.

detonation ■ Overpressurization of an environment when the energy moves at speeds faster than the speed of sound and thereby causes mechanical damage. Backdrafts and explosions are examples of detonation.

deflagration ■ Overpressurization of an environment when the energy moves at less than the speed of sound. A room involved in a flashover will cause a deflagration in most cases.

WEATHER-CAUSED MECHANISMS

Weather attacks structural integrity in many ways. A flood or earthquake can upset foundations, shifting loads from compression and axial to shear or tension and concentric. Or a foundation can be shifted laterally so that it no longer supports any of the loads above.

Snow and rain can impose excessive loads if not properly dispersed. Heavy rain followed by a freeze can have catastrophic results. Wind will often attack a structure laterally as well as from above. The creation of a vacuum on one side of a wall can cause that wall to fail. This concept led to the belief that if you open your windows during a tornado, the tornado is less likely to pull your roof off, but this has been proven to be untrue. In hurricane- and tornado-prone areas the use of stronger connections between foundations and walls, and roofs and walls, is called for. Floods will cause erosion around the foundation and dislodge the structure, causing it to fail. Earthquakes have catastrophic effects on unreinforced masonry structures. This was very graphically demonstrated in Haiti in 2010 when almost all of the buildings in Port Au Prince were destroyed.

Types of Collapses

Failures or collapses occur when the configuration of materials or systems is changed in such a way that they can no longer withstand gravity. The major types of collapse occur in floors, walls, and roofs. Wall failures are subdivided into masonry failures and wood failures.

How Floors Collapse

- Pancake
- V-shaped
- Lean-to
- Tent

PANCAKE COLLAPSE

A **pancake collapse** occurs when a total floor system fails (see Figure 9-2). The weight of the system is transferred to the floor below, which in turn fails. This process may continue all the way to the ground level or it may stop. If the failure stops at a point above the ground floor, the dangers for personnel are heightened. A secondary failure can occur, trapping personnel below and injuring or killing them. This type of failure can occur in wood and masonry floor systems. The failure of the World Trade Center towers on September 11, 2001 began with a pancake collapse on the affected floors and continued downward.

V-SHAPED COLLAPSE

A **V-shaped collapse** occurs when floor beam or joist fails in the middle of a floor (see Figure 9-3). The center of the member(s) falls, while the terminal ends remain in contact with the walls. The **voids** created at these ends provide reasonable expectation for victim survival. This type of collapse is most often seen with wood floors, but can occur with concrete floors placed on Q decking.

LEAN-TO COLLAPSE

A **lean-to collapse** is the result of the failure of one end of a joist or beam (see Figure 9-4). The affected end falls and the result is a void between the wall surface and the upright portion of the floor.

TENT COLLAPSE

Tent collapse is the opposite of a V-shaped collapse (see Figure 9-5). In a tent collapse the terminal ends of the member fail and the center of the floor or roof assembly snaps, which results in the tent-shaped failure. The void under the center may contain victims. This type of collapse primarily occurs in wood systems.

pancake collapse ▪ Complete failure of a roof or floor support assembly. The resultant collapse falls directly down on the floor below. This total failure can trigger subsequent collapses of supporting lower floors until total failure of the structure occurs.

V-shaped collapse ▪ Failure of a roof or floor assembly in which the center supports fail and the roof or floor members subsequently fail in a V-shape in the middle of the span. This is the opposite of a tent collapse in which the end supports fail and the center of the floor snaps up resulting in a tent shape.

voids ▪ Spaces that occur due to falling debris from failing assemblies stacking up in such a fashion as to create small tunnels or cavelike areas within the debris field of the failed structure.

lean-to collapse ▪ Failure of one end of a supported wooden floor. The remaining end continues to be supported.

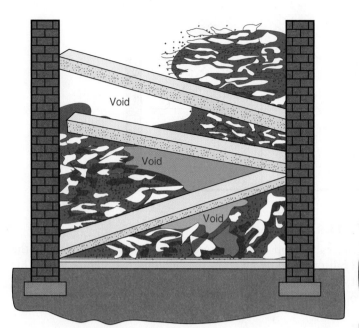

FIGURE 9-2 Pancake collapse. Consider weight of all debris and length of time for rescue operation.

FIGURE 9-3 V-shaped collapse. Victims may survive in voids created at the base of the wall. Breach wall at this point and use caution.

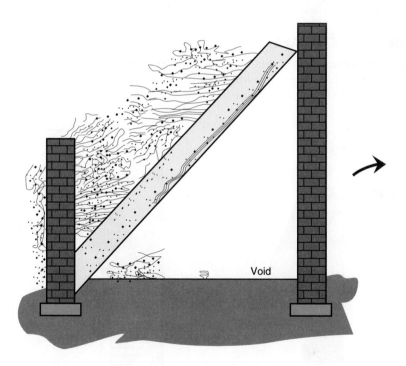

FIGURE 9-4 Lean-to collapse. Be wary of secondary collapses as floor and debris try to push the wall outward.

How Masonry Walls Collapse

- Inward/outward collapse
- 90-degree outward collapse
- Curtain wall collapse
- Parapet wall collapse

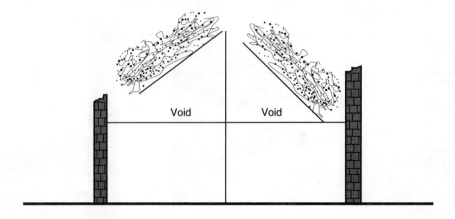

FIGURE 9-5 Tent collapse. Common in wood frame (Type V) with center support.

Void   Void

INWARD/OUTWARD COLLAPSE

inward/outward collapse ■ Collapse of masonry wall at the intersection of floors. The top section of the wall falls in toward the structure and the bottom kicks out away from the building. This can also affect wooden walls with masonry veneers.

Inward/outward collapse causes the top of a wall section to fall inward towards the building and the bottom to fall outward away from the building (see Figure 9-6). This collapse is usually the result of lateral impact from within the affected structure or from a heat-caused failure of a component along the longitudinal section of the wall.

90-DEGREE COLLAPSE

90-degree collapse ■ Total failure of a wall that falls straight out from a building. When collapses of this type occur in masonry walls, pieces of masonry can bounce or fly through the air, causing injury or death.

90-degree collapse occurs when the lower portion of a wall fails and the entire wall falls away from the building (see Figure 9-7). Pieces of broken masonry will often continue to move laterally after the wall has made contact with the ground.

CURTAIN WALL COLLAPSE

curtain wall collapse ■ Occurs when masonry is shifted from compression to tension. The masonry members (brick and blocks) fall straight down in the same fashion as a curtain falls when released from its supports.

Curtain wall collapse often occurs as a result of the impact of hose streams on a clay brick wall or from a ladder truck striking a wall during operations. The bricks fall straight down, causing a rubble pile at the base of the structure (see Figure 9-8).

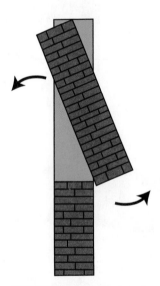

FIGURE 9-6 Inward/outward collapse (masonry). The weight of masonry falling inwards will cause secondary collapse outward.

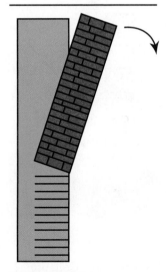

FIGURE 9-7 90-degree collapse. Keep all firefighters beyond collapse zones. This includes personnel operating tower ladders or buckets.

PARAPET WALL COLLAPSE

A parapet wall is a free-standing masonry wall above the roof line. This type of wall collapse occurs as forces within the confines of the roof framing change or when there is a loss of stability along the supporting walls. A parapet wall collapse can be any of the above-mentioned types.

How Wooden Walls Collapse

- Lean-over collapse
- 90-degree collapse
- Inward/outward collapse

LEAN-OVER COLLAPSE

This type of collapse, which occurs in wooden walls, is usually associated with serious weather or fire damage that has caused a structure to come apart, at which point the weight of the total structure folds over to one side much like an empty cardboard box might (see Figure 9-9).

90-DEGREE COLLAPSE (WOOD)

This type of wooden wall collapse is the result of weather or fire impacting one side of a building. The connections are lost and the wall simply falls over to the ground (see Figure 9-10).

INWARD/OUTWARD COLLAPSE (WOOD)

An inward/outward collapse is the most dangerous of all wall collapses. It occurs without warning and can affect all sides of a building. Masonry walls most often fail individually, but when a wood wall suffers this type collapse it can affect all floors and sides. It can easily cause pancake collapses of the floor systems within the structure (see Figure 9-11).

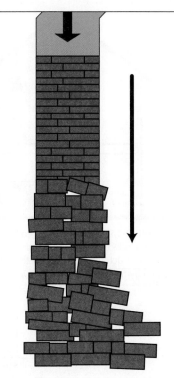

FIGURE 9-8 Curtain wall collapse. This type of collapse occurs when masonry simply drops in place. Often caused by hose streams.

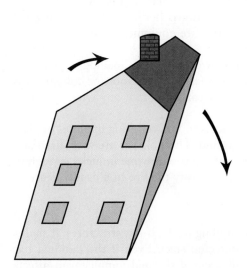

FIGURE 9-9 Lean-over collapse (wood). This is caused by wind or fire in wooden structures.

FIGURE 9-10 90-degree collapse (wood). This is common in balloon-framed wooden structures.

FIGURE 9-11 Inward/outward collapse (wood). This occurs in trussed or conventionally framed wooden structures when connections are impacted by fire at intersections of wall and floors.

Collapse Potentials of Buildings By Type

TYPE I: FIRE-RESISTIVE

Type I construction has the best record of allowing firefighters to operate at a scene without major collapse. This type of construction is widely used for commercial buildings housing large numbers of people, such as hotels, offices, churches, or schools. The buildings can be low-, mid-, or high-rises. They often occupy a large footprint of 20,000 square feet or more. The primary firefighting issues related to these buildings are that they lack sprinklers and smoke travels rapidly through them.

When major fires do occur in these buildings the effects of heat on concrete will cause safety hazards for personnel operating on the scene. If sprinklers are absent, if they malfunction, or if the fire overpowers the system then fire personnel will be impacted by small but potentially deadly failures.

Some common failures that occur with this type of construction are:

- Glazing failure from thermal impingement
- Floor heaving: uplift of Q decking from thermal effects
- Floor sagging: Q decking losing tensile strength from thermal effects
- Suspended ceiling collapse from thermal impingement
- Elevator failure from effects of thermal and smoke conditions
- Spalling of concrete

Size-up

The first units to arrive at a Type I structure must determine the volume of fire and how the fire is impacting the main structural components. Because these buildings are often public assemblies, the fire attack is delayed while rescues and evacuations begin. While the mass of the building's components allows for lengthier thermal impingement before failures begin, the size or height or a combination of both makes these structures very dangerous to operate in. If the building is not sprinklered and the fire involves more than several rooms, then large-diameter attack lines are called specifically for. But in all instances, the need for larger lines to support the original attack lines is indicated.

The location of the fire can have broad ramifications. If the fire is attacking the machine rooms or the electrical vault, failures of the fire equipment and the elevators may certainly occur.

Spalling of concrete will occur if the temperature exceeds 500°F. Spalling is the explosive propulsion of the top layer of cement from the affected wall, floor, or column. It is to be anticipated that suspended ceilings will fail very early and that any wires or cables supported above this system will also fail.

Fires on upper stories can cause the glazing systems to fail, resulting in pieces of glass that sail for long distances before they strike the ground. These shards are very sharp and can cause severe lacerations, amputations, or decapitations of anyone unlucky enough to be struck. The influx of air from failed glazing can intensify the fire to a deadly degree.

TYPE II: NONCOMBUSTIBLE

Type II construction is the predominant form of building for high-rise structures. It shares many similarities with Type I but contains unprotected steel. Even if the building has sprinklers, failures of major components will still occur if thermal impingement affects the steel members.

Many of the newer commercial assemblies are composed of lightweight steel trusses and lightweight metal skin. The roof may be Q decking, steel purlins supporting a metal roof deck, or fire-treated plywood. All these components will be affected by heat.

Some common failures associated with this type of structure are:

- Failures of roof supports
- Floor failures
- Column or beam failures
- Exterior wall failures

When temperatures reach 1,100°F the steel members begin to elongate. Their movement will lead to column failures and wall failures as the steel pushes against the restraints. When temperatures approach 1,700°F, the steel begins to lose its ability to remain upright and sagging occurs. After the temperature reaches 2,000°F the steel begins to decompose completely.

As these failures occur firefighters are exposed to heavy falling debris, which can result in firefighter injury or death.

Size-up

The exteriors of these mid-rise and low-rise buildings often contain metal. The walls, parapets, and roofs may be metal or masonry. The most reliable method for determining Type II construction materials is to open up a ceiling and look for unprotected steel. Water must be applied quickly to steel members impacted by heat or flame.

TYPE III: ORDINARY

Type III construction is common in buildings that date from the nineteenth century and can be found in residential and commercial occupancies.

Some common failures associated with this type of construction are:

- Partial collapse
- Parapet wall collapse
- 90-degree masonry collapse
- Stair collapse
- Floor collapse: pancake, V, lean-to, or tent
- Cornice failure

The effects of hose streams on mortar can cause failures as the force of the stream washes out the lime/sand heavy mix. If the building configuration is row style in which units share party walls, and if the building lacks a lateral partner, then **partial collapse** or total collapse can occur.

Parapet walls are particularly vulnerable when fire is attacking the cockloft area or impinging on the underside of the top floor. A parapet wall can fail totally or individual pieces of the coping or clay masonry can fall. In either event firefighters under the failure will be impacted.

Cornices are found at the intersection of the vertical walls and the roof. They can be constructed of light metal (tin), masonry, or wood. The masonry is attached by corbeling out from the vertical walls. Wood or tin, however, may be attached in a variety of methods. Sometimes wooden nailing blocks take the place of brick or stone; the cornice assembly is then nailed into the blocking resulting in a hazardous shear assembly. Another approach is to attach the cornice by cast iron straps passed through the masonry; this option is extremely vulnerable when fire on the inside attacks the cast iron. The key is to remember that cornices are above you and they will fail. Look up as you perform size-up and include them in your risk assessment.

partial collapse ■ The less than total failure of any structural member, roof, or subassembly (doors or stairs).

The interior may be vulnerable to failures because of the age of the components. Stairs, in particular, will fail quickly when impinged by thermal conditions. Ceilings and walls, which contain lathe and plaster, can fall, and injure fire personnel.

Size-up

The key for Type III buildings is to pay attention to architectural features such as the elaborate masonry work on the front, parapet walls, and turrets or towers common in the end units of the row. The buildings are often long and narrow, under 20′ wide and 60′ to 80′ long. The throats of the doorways and windows should at least 8″ deep. The stairs will be stacked. Be cautious of recently constructed units designed to imitate Type III styles, but that are actually Type V with veneers.

Firefighters should use extreme caution when structures in a row configuration are missing their lateral partners. Fires on multiple floors, or significant fires on lower floors or roof areas, will lead to partial or complete failures of exterior or parapet walls. Limit interior operations to lifesaving and establish collapse zones quickly.

TYPE IV: HEAVY TIMBER

Type IV buildings are the earliest forms of mid-rise and high-rise structures that have an internal skeleton. Many of these structures began their lives as manufacturing centers, churches, or schools, but have been converted to multifamily, office, or other public assembly uses. Many still contain early nineteenth-century trusses, which use cast iron connectors or turnbuckles.

Some common collapse failures associated with this type of structure are:

- Masonry wall collapse
- Floor failures
- Ceiling collapse
- Roof collapse
- Interior wall collapse

The size of the main structural members, 8″ × 8″, can lead to a false sense of security. The fire does take longer to take hold in these structures, but once it does the intense volume of heat can easily overcome the fire department's abilities and resources. Fire intensity can be a by-product of many decades of oil and other chemicals soaking into the tongue-and-groove floor systems.

Firefighter injury or death at these structures occurs because either they or the incident commander did not respect **collapse zones**. If a significant fire is suspected or confirmed, then collapse zones must be established. All areas within the zone must be restricted for essential activities only and then only for a short time. As noted above, although the 8″ × 8″ main structural members can lead to a false sense of security, and although it does take longer for the fire to take hold in these structures, the eventual intense volume of heat can easily overcome the fire department's abilities and resources. Remember, fire intensity can be a by-product of many decades of oil and other chemicals soaking into the tongue-and-groove floor systems.

Size-up

The size, use, and age (over 100 years) of these buildings make recognizing them easy. The existence of buttresses is another good indicator. For buildings that have been converted to another use, caution must be used because the voids created can lead to significant fire development. The use of large-caliber streams, while anticipating of collapse, is really the only safe method by which to combat these giants.

collapse zones ■ Restricted areas surrounding a building that has been damaged by fire or natural disaster and is likely to collapse as a whole or in part. Firefighters and apparatus should not be placed within the confines of this zone because of the danger of falling debris. The standard distance out from the building is one and a half to two times the height of the building.

TYPE V: WOOD FRAME

Type V structures are the sites of most structural incidents. Failures of their components can result in significant dangers to responding personnel if the personnel are unable to recognize the point at which heat development is sufficient to cause failures to main assemblies of the structure. This is especially true if the building contains trusses (floor or roof or both), TGI beams, or microlam. The use of SIPs and OVE will also lead to collapses but their use is so new data is not yet readily available.

Some of the more common failures associated with this type of structure are:

- Roof collapse
- Floor collapse
- Stair collapse
- Interior wall collapse
- Masonry veneer failure

For buildings that contain trusses or other manufactured assemblies, the responding personnel are racing against failures as soon as they arrive on the scene. When the temperature reaches the point at which connection points fail, the truss and other manufactured assemblies will come apart and a collapse will occur. With the failure of a single truss, personnel should expect the failure of others and anticipate the possibility of a total failure of the assembly. For buildings that contain SIPs, if the flame and heat penetrate the outer skin, the foam interior will rapidly contribute to the fire load and cause the structural instability of the panel. The problem with QVE framing is the lack of redundant framing so when structural members become impacted the subsequent load shift will be catastrophic.

For buildings constructed of balloon framing, the danger lies in the fact that many of the floor assemblies in these structures are in shear. For any building constructed of wood, the building will fail if the fire has affected the main assemblies of walls, floors, or the roof. This is the only form of construction that can become fully involved.

 SCHOOL OF HARD KNOCKS

I have been a student of building construction throughout my years in the fire service, starting with my time at the Clifton Heights Fire Company in Clifton Heights, Pennsylvania, and continuing until my retirement from the District of Columbia Fire Department.

I was fortunate to have Deputy Chief Vinny Dunn, FDNY, as my mentor. In my opinion, he is the preeminent expert regarding building collapses as they relate to the fire service. He told me the story of the 23rd Street fire fought by the FDNY in 1966, when if it weren't for an order to have his company stand by in front of the structure, he would have joined the eleven other members of the department who perished that night. Every firefighter and commander should read the account of the 23rd street fire to better prepare for fire combat.

I cannot tell you how many times his story has helped me prevent my firefighters from being injured or killed on the scene. We, as incident commanders, have to understand collapse potential and act accordingly to protect those who serve under us. We need to predict when failures may occur and choose the safest approach, preventing failures from impacting our people. Commanders who believe that only a fast and aggressive attack solves problems will probably injure or kill someone during their careers. If incident commanders do not respect truss failure potentials, they are directly responsible for the outcome.

FIRST RESPONSE

Churches

Churches are considered to be places to worship to one's God. They are depicted in movies and books in varieties ranging from simple wooden structures to elaborate stone cathedrals. These realities have changed and the fire service needs to understand the changes.

Until the mid 1960s, the church was often the social hub of many urban, suburban, and rural areas. In urban areas, great edifices had arisen during the eighteenth and nineteenth centuries. These slate-roofed cathedrals of stone and stained glass often exemplified the prosperity of a neighborhood. In suburban areas, newer buildings were made of masonry, with plain windows and asphalt or fiberglass-shingled roofs, and indicated the start of a religious community. In rural areas, wooden churches of painted white clapboard with slate-shingled roofs indicated stability.

Churches today are found in many different settings, even mall storefronts. Modern cathedrals are most likely made from stone veneer, with manufactured roof members and imitation slate on the roof, and, depending on the code, steel structural skeletons.

The role of the church has also changed. Churches are no longer occupied solely on weekends. Today's churches serve many purposes, including providing space for schools, shelters, and community meeting, and may be occupied around the clock.

Most churches, except those in storefronts, share one important feature relative to planning fire service strategies and tactics. The voluminous sanctuary often is so high that hand-held 1½" and 1¾" handlines cannot reach the ceiling. In the 1970s a Montgomery County, Maryland fire chief, Chief David B. Gratz, came up with a tactic he called the blitzkrieg for churches. His plan was to deploy unmanned monitor nozzles or deluge guns to put out church fires, especially fires involving the ceiling or the bell tower, right from the beginning of the incident. Forget the exercise in futility that is the attempt to place attack lines in service only to be driven out eventually and still lose the building.

I thought his idea had merit then and I still believe so today. Most churches carry insurance. If the structure is vacant or unoccupied at the time of the fire, the question must be posed, what is to be accomplished? Units are dispatched to put the fire out and make rescues. If there are no rescues to be made, only the job of putting out the fire remains. There are tools on the apparatus that will accomplish this task without risking the lives and livelihoods of the responders. It takes fewer personnel to deploy a master stream, which can be completely effective, than to deploy handlines. Older churches may contain trusses made from cast iron and wood while the newer churches may be constructed only from manufactured structural members. Both of these assemblies pose dangers to firefighters. The footprint of most churches is such that when visibility is poor or nonexistent, it is extremely easy to become lost and disoriented. Only pre-planning, training, and drills prepare the firefighters to fight fires in these occupancies.

Some strategical and tactical considerations are listed below:

- All churches should be pre-planned as they are certainly target hazards.
- All responding units should have an available copy of the floor plan.
- Collapse zones must be established from the onset of the incident.
- Additional safety staff, having building and fire service experience, should be deployed to monitor conditions on all sides of the building.
- If roof ventilation is necessary, keep the pitch of the roof in mind. Pitches above 6/12 are unworkable and unsafe. Dislodged slate tiles can be deadly.
- Stained glass windows can be replaced. If horizontal ventilation is needed, the windows must be sacrificed.
- Interior offensive attacks are called for only if searches or rescues are mandated. Information from a credible source is necessary before commencing such operations.
- Consider placing ladder/truck companies at the corners of the structure from the beginning so that they can deploy ladder pipes without relocating apparatus.
- Incident commanders should always err on the side of the safety of their firefighters.

Your unit has been dispatched to the scene of a fire in a four-story multitenant Type III constructed building. The affected building is in the middle of a row. The fire is on the top floor and involves three to four rooms in the front half of the building. There are occupants still exiting the building when you arrive. There appear to be pieces of broken masonry on the sidewalk in front of the building.

1. What hazards will confront you and your company during this incident?
2. What information must you relay to the incident commander?
3. What actions can you take to protect yourself and your company?

Summary

Understanding the mechanisms of collapse is essential to firefighting safety. Structures tend to collapse in predictable ways, so it is important to know how to size them up according to their type (I–V) of construction. Collapse is affected by many factors, including weather, human error, and fire. Residential construction, which suffers more fires than commercial construction, is subject to less scrutiny than commercial construction and it is therefore likely to have faulty workmanship. Remodeling can impact the original support system in detrimental ways. Be aware of all of these factors when conducting size-ups.

The thermal effects of fire can cause a building's structural system to give way, shifting loads and causing collapse. There are many kinds of collapse; it is essential for incident commanders to familiarize themselves with all of these and to establish collapse zones to keep firefighters safe while they work. It is important to understand that victims may be trapped in voids in collapse zones. Identifying the type of collapse will help focus the rescue operations. Keeping people safe is paramount and predicting a collapse is less difficult if you know and understand the events that lead to collapse.

Review Questions

1. Discuss how wood frame buildings fail.
2. Would you expect to find survivors in a four-story Type I or Type II with pancake collapse of all floors?
3. What are the dangers related to parapet walls?
4. Discuss collapse zones and their relevance to Type IV construction.
5. Discuss how steel fails.

Suggested Reading

Avillo, A. 2002. *Fireground Strategies*. Saddlebrook, NJ: PennWell Books.

Dunn, V. 1988. *Collapse of Burning Buildings*. Saddlebrook, NJ: PennWell Books.

Fire Fighter Fatality Investigation Report F2004-17. 2006. Morgantown, WV: National Institute for Occupational Safety and Health.

NFPA 1620. 2003. *Recommended Practice for Pre-Incident Planning*. Quincy, MA: National Fire Protection Association.

NFPA 1670. 2004. *Operations and Training for Technical Search and Rescue Incidents*. Quincy, MA: National Fire Protection Association.

10

Command of Collapse Incidents

OBJECTIVES

After reading this chapter, you should be able to:

- Define five operational phases of structural collapse response.
- Identify hazards and conditions associated with a structural collapse.
- Identify factors associated with rapid scene assessment.
- Define various levels of capability for a structural collapse incident.
- Describe operational elements within the command structure.
- Identify, request, and apply specialized urban search and rescue (US&R) resources.

Resource Central

For additional review and practice tests, visit **www.bradybooks.com** and click on Resource Central to access book-specific resources for this text! To access Resource Central, follow directions on the Student Access Card provided with this text. If there is no card, go to **www.bradybooks.com** and follow the Resource Central link to Buy Access from there.

Introduction

Structural collapse incidents are considered low-probability but high-consequence events. Structural collapses may be caused by a variety of things including construction accidents; structural deterioration; fire or explosion; natural events, such as hurricanes, tornadoes, floods, and landslides; and transportation accidents. These incidents pose special and unique dangers to responders, including confined spaces, flammable or toxic hazards, oxygen-deficient atmospheres, and unstable rescue areas creating the possibility of secondary collapse. The incident commander and all other responders, for these events, must recognize early on the total complexity and magnitude of the event and begin the process of command, which includes requesting adequate resources and sufficient personnel to assist in mitigating the incident safely.

 FIREHOUSE DISCUSSION

The Alfred P. Murrah Federal Building was located in Oklahoma City, Oklahoma. It was subjected to a truck bomb on April 19, 1995. The blast resulted in 168 fatalities and 680 individuals being injured. The fatalities included one responder, a nurse, who was struck on the head by falling debris during the incipient period of the response. The blast damaged 324 buildings within a sixteen block radius. This was the second bombing of a building in the United States. The earlier bombing had occured in the North Tower of the World Trade Center in New York City on February 26, 1993.

The Building

The Murrah building was a nine story high-rise structure. It measured 220′ × 100′ on each floor. The nine story main building was flanked on either side of the first floor by one story ancillary structures. The roof of the building was constructed of reinforced concrete slabs; concrete columns supported the floors. The exterior sides of the building were constructed of granite curtain walls; there were 5′ × 10′ panels of glass in an extruded aluminum frame in the front. The rear of the building consisted of smaller windows and granite panels supported by spandrel beams. The front façade, containing the glass curtain walls, was also supported by spandrel beams. These beams were resting on three columns. The four corners of the building contained hollow columns, which carried the HVAC exhaust and intake shafts. The core area was in the southeast corner of the building and contained the stairwells, electrical services, plumbing services, and elevators.

The Bomb

The bomb consisted of 4,800 pounds of ammonium nitrate fertilizer, nitromethane, and diesel fuel. The bomb was constructed of (13)55 gallon drums. Nine of the drums contained nitromethane and ammonium nitrate. The other four contained fertilizer and 4 gallons of diesel fuel. This type of bomb is known as a binary explosive, which is an explosive consisting of two nonexplosive components that become explosive when combined. The fuse system consisted of shock tube, which is a nonelectric explosive initiator that is safer to handle than a detonator cord. The shock tube ran to 350 pounds of Tovex Blastrite sausages. Tovex is a water gel explosive composed of ammonium nitrate and methylamonium nitrate. It has widely replaced nitroglycerin as a blasting agent.

In comparison, the 1993 bomb used at the World Trade Center consisted of 1,310 pounds of urea nitrate surrounded by aluminum, magnesium, and ferric oxide. Three cylinders of hydrogen were placed around the bomb to enhance the force. The fuse was a simple powder-covered initiator. The bomb was placed inside of a van and parked in the basement of the tower.

The Bombing

At 0902 hours on April 19, 1995 the Murrah Building bomb was detonated. The blast obliterated three of the main columns in the front of the structure, which in turn dislodged the spandrel beam on the third floor causing a pancake collapse of the front third of the building from ground level to the roof. Buildings across the street were heavily damaged. Cars parked in the lot across from the building were instantly set on fire and secondary explosions began occurring as their gas tanks began erupting (see Figure 10-1).

FIGURE 10-1 Effects of bombing at Alfred P. Murrah Federal Building. *FEMA News Photo.*

The Response

The blast could be heard forty to fifty miles away from the building. Almost immediately, responder units from inside the city as well as the surrounding environs began to self-dispatch. When the first units arrived on the scene, they were met by a scene reminiscent of Dante's Inferno. Many hundreds of dazed, injured, and confused victims milled aimlessly in the blocks surrounding the blast; there was total chaos and panic.

The members of the Oklahoma City Fire Department, assisted by other members of the Public Safety Force and volunteers, went into action. It was during this period that a nurse was killed by falling debris. During the first hour, fifty victims were rescued from the building site. The state emergency operations center was set up and included representatives of many of the state agencies.

A total of 12,000 personnel from many agencies at the local, state, and federal levels responded to the disaster. The Federal Emergency Management Agency (FEMA) dispatched eleven of its urban search and rescue (USAR) teams, providing 650 highly trained and competent technical rescuers.

The chaos of the first hour, when all units were operating independently, was alleviated at 1028 hours when there was a report of a second bomb and all responders were ordered to evacuate the building. It was during this period that the incident commanders and their senior staff were able to implement fully the incident command system. Common terminology was essential to clear communications among the many arriving agencies. The respondents were divided into branches, divisions, and groups. A response of this magnitude requires the use of branches immediately upon the arrival of assets.

After the 1993 World Trade Center bombing the Fire Department of New York (FDNY) did not request mutual aid from the surrounding areas. All bombing attacks are law enforcement responses, therefore there will be many agencies from the state and federal government to coordinate and cooperate with. Oklahoma City's responders did a truly outstanding job at incorporating all of the supporting and assisting agencies. The rescue efforts ceased on May 5, 1995. All but three of the victim's bodies had been removed. The remainder of the building was imploded on May 23, 1995.

The main contributing factors to the damage and destruction caused by the Murrah Building bomb were the concrete building materials, the window assemblies, and the spandrel beam configuration. The concrete of the building lacked ductility. It was not flexible enough to allow for the bending caused by the blast. Concrete is excellent in compression and terribly weak in tension. In the late 1980s, efforts to make concrete more ductile were highly successful. The product and process was developed at the University of Michigan. This engineered cementitious composite demonstrates the same ductility properties as steel. The coarse aggregates normally put into concrete mixtures are replaced with sand, fly ash, and microfibers. This allows the concrete to bend; a secondary benefit is that the concrete is lighter in overall weight.

Another problem related to the concrete was the attachment mechanisms at the columns. The slab lost contact with the column at the moment of overpressurization. The slab was lifted up (in tension) and when it rebounded, the connection was broken. (Note: The construction industry later addressed this problem through the use of stirrups, rebar bent into a loop, attached to the column rod work, and poured as one integral assembly.)

The windows could not withstand the overpressurization caused by the pressure wave of the explosion. The glazing disintegrated, causing secondary damage from flying shrapnel. The aluminum frames could not restrain the glass. (Note: Correcting this defect involves the use of laminated glass. This glass will have a polyethelene sheet installed while the glass is being manufactured, resulting in a product similar to car windshields but stronger. The frames are also substantially strengthened and include the use of silicon sealants to retain the glass.)

As a result of the Murrah bombing and subsequent terrorist attacks, the federal government has incorporated these enhancements in its specifications for new federal buildings.

 THINK ABOUT IT!

Not many fire departments have the resources of Oklahoma City or New York, which makes it important that they have mutual aid plans in place and understand how the federal response is structured. Combined training and exercises involving the various agencies and departments should be implemented as should a common communications plan with central channels available so that all responders can talk to one another. The incident commanders must possess the skills and knowledge necessary to command an operation this large and complex. These types of responses require more than just utilizing the incident command system.

- Discuss the many complex issues facing the first arriving units and how might the issues be prioritized.
- Discuss how weather would play a role.
- Discuss how to control civilian volunteers.
- Look at the various buildings in your area that might be targets for domestic or international terrorism. Determine what steps are being taken to provide them with greater security.
- Discuss the role of NFPA 1670 and the impact it has on the initial responders.

Rapid Scene Assessment

ROLE OF FIRST RESPONDERS

The first arriving units at the scene of a structural collapse must perform a complete size-up. They must also change their behavior paradigms from aggressive and emotional to cautious. They must perform structural triage, which helps to identify, select, and prioritize the structures in which there is the highest probability of success with respect to finding and rescuing live victims. Structural triage is accomplished using the following steps:

- Obtain pre-collapse intelligence. This includes information from witnesses and victims, building diagrams or plans, and occupancy information.
- Deploy reconnaissance teams to evaluate structural conditions, hazards, and rescue opportunities (this may require the services of a structural specialist and HazMat specialist). This information assists in determining hazard versus risk in **rescue operations.**
- Analyze information and determine the rescue risk to benefit ratio. Some hazards will prompt a no-go until the hazard is mitigated. This may occur even though victims are known to be in the area.

rescue operations ■ Period of time when all efforts are at maximum to attempt viable victim extrication. The use of hand equipment is called for to prevent injury to victims.

- Prioritize rescue sites. The highest priority sites are those where the most victims can be rescued safely in the shortest amount of time.
- Reevaluate continually. All relevant information must be processed by command. As conditions change, strategies and tactics may change as well.

OPERATIONAL PHASES

There are five operational phases associated with structural collapse incidents.

- Phase I: Initial response
- Phase II: Expanded (reinforced) response
- Phase III: Extended response (twenty-four hour operations)
- Phase IV: Demobilization
- Phase V: Return to a state of readiness

Phase I: Initial Response

Phase I is the most critical period during a response to a structural collapse. The initial minutes of a major structural collapse are pivotal to a successful and safe conclusion. The first units on the scene must recognize the potential for escalation and begin to request additional resources. The need to gather information is paramount. How many victims were in the building at the time of the failure? Where are they? What was the original configuration of the building compared to where it is now? The need to ascertain the integrity of the remaining portion of the building or buildings is crucial before rescues can be attempted. The first arriving officer must establish command and begin to use the National Incident Management System (NIMS). Responses of this magnitude require the use of branches with prompt support by divisions and groups. Many personnel will be needed. For many smaller departments this will necessitate the use of mutual aid; all responding units must be clear as to mission and accountability. Use common terminology understood by all units and other agencies that will arrive later in the incident.

The initial units must begin to clear the area of apparatus because the vibrations and noise of the engines may preclude hearing cries for help. The vibrations may also contribute to further failures of the building. The first arriving units must survey the total site footprint for potential or existing hazards, including, for example, hazardous materials, live wires, or leaking natural gas lines. Each hazard must be marked and identified.

At this stage any rescue attempts should be conducted without entering the failed structure or the debris field. The use of aerial ladders or tower ladders is called for in order to prevent death or injury to responders; the building may still be failing. The possibilities for secondary collapse during the first minutes are many because stabilization has not yet taken place. One of the first priorities of the incident commander is to gain control of the responders and collect information so that appropriate tactical and strategical decisions can be made.

Phase II: Expanded (reinforced) Response

It is during Phase II that senior command officers begin to arrive. Before assuming command, it is imperative that they perform their own size-up of the situation, including a 360-degree walk around the scene. The **transfer of command** should take place face to face. The briefing should include the progress of scene setup, the deployment of personnel (as indicated on the incident command system chart), identified hazards and progress towards their mitigation, and any other information essential for establishing command.

The arriving incident commander must develop a risk management plan and appoint people to staff positions. The plan must take into account the potential for injury or death to rescuers if they attempt non-surface rescues. Extrication operations can be quite

transfer of command ■ ICS term. The assumption of command and all attendant responsibilities from a junior officer by a senior officer. Usually performed face to face.

complex and lengthy. The incident commander must evaluate multiple victim locations and determine the priority for attempting rescue operations. The go-no-go authority and responsibility rests solely with the incident commander. If the extrication is too dangerous or the victim may not be viable at the conclusion of the extrication, then the resources must be deployed elsewhere at the scene. If additional resources have not been ordered, this must be done immediately to avoid any further delay caused by the reflex time between requesting resources and their deployment on the scene. If additional resources have been ordered, the IC must evaluate the progress being made and determine if yet more resources are required. The IC must also be conscious of the time of day, time of year, and anticipated weather over the next twelve hours. This evaluation will allow the IC to order additional resources such as lights, generators, heaters, lumber, and so on. The logistics position will be discussed later in the chapter, but it would be a good practice to staff that position during Phase II.

The safety officer (SO) position must be staffed and given enough personnel to support the operation. These personnel will be called assistant safety officers (ASOs). The SO must have full authority to order operations to cease if deemed to be unsafe.

As resources begin to arrive, it is extremely beneficial to create a **choke point** and a **staging** area. A choke point is an area away from the incident site to which all apparatus and personnel report. With these types of incidents, personnel, as opposed to apparatus, is what is needed. The apparatus can be parked at the choke point and the personnel transported by bus, for example, to the staging area. Staging is the first check-in position used to monitor accountability. All resources should report to the staging manager and that officer should begin to sort out resources based on use and sequence of deployment. For example, the technical rescue personnel (and their apparatus) and emergency medical personnel will be deployed sooner than will firefighters.

Phase III: Extended Response (twenty-four hour operations)

For larger events, Phase III is the period of time during which required resources should be on-scene and all operations underway. The incident may be at a single building or at many structures spread out over a large footprint. The need to divide resources into work cycles, usually eight to twelve hour shifts, occurs during this period.

The need for documentation is heightened during the work cycles. Each successive team evaluates the progress made and then creates the incident action plan (IAP) for the following shift. The planning position, which will be discussed later in this chapter, must be staffed by this time.

The decision to transition from rescue to **recovery operations** is made during Phase III. The rescue phase involves all resources actively engaged in trying to rescue and remove victims from the collapse. Great care is taken to prevent further injury to victims so hand tools are utilized. Recovery begins when it is determined that possibility of finding viable victims no longer exists. Heavy equipment is used in recovery; great care must be exercised to prevent the public from viewing what may be a macabre scene.

Phase IV: Demobilization

It is during Phase IV that units begin to disengage from the scene and return to service. Evaluations must be performed to ensure that the process occurs safely and efficiently. Many personnel are injured during this phase so safety must continue to be a prime focus. A plan for demobilization should be initiated by the planning section chief in cooperation with the operations chief upon approval of the incident commander in charge.

Phase V: Return to a State of Readiness

These types of events are physically and psychologically taxing for all personnel. **Critical incident stress management (CISM)** teams will accelerate responders' ability to return to normal faster. A post incident analysis should be convened and lessons learned from activities accomplished.

choke point ■ Area away from the scene where all responding resources are forced to stop and begin accountability. This can be a parking lot or other large area. Apparatus can be stored here. Security will have to be provided.

staging ■ Area close to the scene from which all resources are deployed. Units who have been sent to rehab will report back to this location for reassignment. In high-rise incidents, this area is two floors below the fire.

recovery operations ■ Period of time after all viable victims are assumed to have already been removed and the process of body recovery begins.

critical incident stress management (CISM) ■ Team of counselors and other professionals designated to address psychological problems of responders associated with incidents.

COMMAND STRUCTURE

Incidents of this type will overwhelm most departments and municipalities. They often exceed pre-plans established for fires. The use of NIMS is a prerequisite for a safe and effective outcome, however, NIMS, or ICS as it is commonly referred to, requires careful consideration, not just assigning bodies to fill in boxes on an organizational chart. A qualified and experienced individual should be assigned to each box on the chart. That person may come from another agency or department; too often smaller departments feel compelled to complete the exercise with in-house staff. This insecurity can result in commanders from other departments lacking confidence in the system.

As part of the size-up, the incident commander must recognize the potential for the incident to overwhelm the department. The positions within the **command staff** and **general staff** will have to be manned. A **unified command** may have to be established. The deployment of resources, as they become available, has to be managed through the use of groups, divisions, or branches. It is essential that the incident commander have the knowledge and experience to command an incident of this magnitude.

COMMAND STAFF

The command staff positions are key in supporting the IC as s/he works to grasp the big picture. The command staff positions are:

- Safety Officer
- Liaison
- Public Information Officer
- Staging Manager

Safety Officer

The safety officer (SO) must have knowledge and experience as well as maturity. The safety officer must be compliant with following NFPA standards *(NFPA 1521 Fire Department Safety Officer),* and should be knowledgeable, experienced or educated regarding building construction methods and materials. The position must be staffed upon receipt of the alarm. The SO must perform a size-up noting hazards and the integrity of the structure even while rescues are being attempted. The SO must be part of any decisions involving extrications from the collapse site and must provide for air monitoring. The SO works in concert with operations and EMS personnel. Patient care is critical during rescue attempts and evaluations for viability must be conducted before the extrication attempt has begun.

Liaison

This person is the point of contact between the incident commander and representatives from other agencies; s/he must ensure that only those representatives who are legally required have made a resource commitment, or who possess decision-making abilities are allowed into the command post.

Public Information Officer

This critical position is the major connection between the public and the responding personnel. Factual information is essential so that the public will know what is going on. The public information officer (PIO) has to have a good working relationship with the press. All statements issued by the PIO should be approved by the IC, however, as this prevents the dissemination of misinformation that may reflect negatively on the department.

Staging Manager

The staging manager is responsible for coordinating all personnel and apparatus responding to the scene. This person can request additional resources if given the authority by the IC. The staging manager should consolidate resources by type so that an orderly deployment can occur. Remember: the staging area should be far enough away from the scene so as to not interfere with operations.

command staff ▪ ICS term. The positions of liaison, safety, public information, and staging. Positions are staffed to assist the incident commander.

general staff ▪ ICS term. Key positions of planning, logistics, finance/administration, and operations are assigned to assist incident command with major components of the command structure.

unified command ▪ Command matrix including commanding officials from all major agencies performing the functions of joint command. Decisions are reached by consensus and one IC speaks for the group.

General Staff

The incident commander and the general staff make up the overall management team. The team is responsible for all of the resources, equipment, and money needed for an incident of this type. The general staff includes:

- Planning
- Logistics
- Finance/administration
- Operations

Planning The planning section is responsible for gathering, assimilating, analyzing, and processing the information needed for effective decision making. The planning chief position requires an individual with collapse and rescue experience. Some of the responsibilities of the planning chief are:

- Evaluate current strategy and plan with the IC.
- Maintain resource status (RESTAT) and personnel accountability.
- Forecast possible outcomes.
- Gather, update, and manage situation status (SITSTAT).
- Use technical assistance as needed.
- With input from operations, recommend changes to IAP, review NIMS for a more thorough study of this important document.
- Evaluate incident organization and span of control.

Logistics The logistics section is the support mechanism for the organization. Logistics provides services and support functions for all organizational components. This section is probably the key to success within the entire operation, as they must anticipate equipment and personnel needs in a dynamic atmosphere. They must be able to identify requirements for specialized equipment not normally associated with fireground operations. An individual with strong construction experience would be an asset as either the chief or deputy. The logistics chief must be augmented with sufficient personnel to accomplish the many tasks required. Support and services handled by logistics include:

- Command post (CP), base, and other facilities
- Transportation
- Supplies
- Equipment maintenance
- Fueling
- Feeding of responders
- Communications
- Medical services and rehabilitation for responders

Finance/Administration The position of finance/administration section chief is staffed during an incident in which there is a specific need for financial or legal services. The responsibilities of the section chief are:

- Procure services and supplies from all sources.
- Document all financial costs of the incident.
- Document for possible recovery of unused services and supplies.
- Document for personnel compensation and injury claims.
- Handle all legal requirements related to the incident.

Operations The operations section chief is responsible for direct management of all tactical activities, tactical priorities, and the safety and welfare of assigned personnel. This position must be staffed with an experienced individual with the ability to delegate.

S/he must fully appreciate the mission of the special operations personnel and, more importantly, how best to support them. This person is responsible for assigning resources to complete tasks necessary for the mitigation of the incident. S/he is responsible for the hands-on portion of the incident and oversees the five phases of a structural collapse:

Phase I: Assessment of the collapse area

- Search the area for possible victims both on the surface and buried
- Evaluate the structure's stability
- Evaluate utilities, including the possible need to shut them down for safety

Phase II: Remove all surface victims as quickly and safely as possible
Phase III: Search and explore all voids and accessible spaces for viable victims
Phase IV: Remove selected debris (using special tools/techniques) if necessary to rescue a victim
Phase V: Remove general debris after all known victims have been removed

The operations chief must have knowledge of building construction materials and methods, a thorough understanding of the logistical needs to support operations, and experience in technical rescue. S/he must assign units to perform various tasks during the five phases of the operational cycle. To accomplish each task and to prevent span of control problems, the chief must assign units to groups, divisions, or branches. Groups are established by task; for example, a rescue group or extrication group removes victims. Divisions are established by geographic responsibility; for example, all activities on the front side of the incident would go to the alpha division. A branch is established when the task or geographic area is larger than that which can be handled by a single division or group. Divisions are responsible for specific geographical areas, while groups move throughout the incident as their tasks dictate. Divisions and groups report to branches, but not to each other.

UNIFIED COMMAND

These incidents are so large and complex that a unified command structure will have to be established. Whenever multiple agencies or multiple jurisdictions are involved unified commands are established. This can involve multiple local elected officials, fire departments, law enforcement, military agencies, and medical personnel. The command decisions are reached by consensus but no commander relinquishes responsibility for protecting his or her people.

Levels of Capability

We have covered the many attributes of building types and construction methods. When responding to major collapse incidents to which urban search and rescue (US&R, pronounced USAR) teams might be called, we need to understand their terminology and procedures.

CONSTRUCTION CLASSIFICATION

US&R teams classify four general types of building construction. They are:

- Light-frame construction
- Heavy wall construction
- Heavy floor construction
- Precast construction

Light-frame Construction

Structures of this type are constructed with a skeletal frame system of wood or light-gauge steel components that provide support to the floor or roof assemblies. Examples include wood-framed structures used for residences, multiple low-rise occupancies, and light commercial occupancies up to four stories in height. Light-gauge steel frame buildings include commercial business and light manufacturing occupancies and facilities.

Heavy Wall Construction

This type of construction utilizes a heavy wall support for the floors and roof assemblies. Occupancies utilizing **tilt-up construction** are typically one to three stories in height and consist of multiple monolithic concrete wall panel assemblies. These structures also use an interdependent girder, column, and beam system for providing lateral wall support of the floor and roof assemblies. Buildings of reinforced and unreinforced masonry construction are included in this group; such buildings range in height from one to six stories. These buildings include all types of occupancies.

Heavy Floor Construction

These structures are built using cast-in-place concrete slab floors and **pre-stressed concrete slabs** or **post-tensioned concrete slabs**. The vertical structural supports are integrated concrete columns, concrete enclosed steel, or steel frame and carry the floor and roof assemblies. This type also includes heavy timber, Type IV, which may use steel rods for reinforcement. Heights for this type range from single story to high-rise. Most Type IV constructed public assembly occupancies fit this category.

Precast Construction

The structures within this type, which include walls, columns, roof decks, and floors, employ modular precast concrete components. These components are field connected when they arrive on site. The connectors are welded or bolted. The use of rebar and wire mesh inside of the components is prevalent. The occupancies of this type include commercial, mercantile, and public assembly. These structures range in height from single story to high-rise.

OPERATIONAL CAPABILITY

Urban search and rescue teams recognize operational capabilities based on their training, certifications, and equipment competencies. The authority having jurisdiction (AHJ) determines what instructional priorities and levels of training are necessary to perform the required tasks. NFPA 1670—*Standard on Operations and Training for the Technical Search and Rescue Incidents* (2010 edition) is referenced by organizations. NFPA 1006—*Standard for Technical Rescuer Professional Qualifications* (2008 edition) is used by individuals seeking instructor or consultant certification. The established levels of certifications and competencies include:

- Awareness
- Operations
- Technician

Awareness Level

Organizations operating at the awareness level must meet all awareness level requirements for confined space rescue. These organizations are limited to visual and verbal searches for victims. They mark areas that have been searched, size up the scene for potential and existing conditions, identify five types of collapse patterns, and locate potential victims.

Operations Level

Organizations operating at the operations level need to meet all awareness level requirements. In addition, they must be capable of hazard recognition, equipment use, and techniques necessary to operate at collapse incidents involving light-frame, ordinary, and unreinforced and reinforced masonry construction.

---

tilt-up construction ▪ Concrete is poured in place on the ground on-site. The resulting walls are then raised into place and stabilized by interconnecting metal bar joists.

pre-stressed concrete slabs ▪ Preliminary stresses are placed in a structure member before a load is applied. Concrete is usually pre-stressed by imbedding a high-strength steel wire in tension in a concrete member.

post-tensioned concrete slabs ▪ Post-tensioned concrete can be either precast or cast in place. Ducts are installed in the slab during the pour. Wires are inserted and drawn into tension by way of a pump, then secured. These cables should never be cut or allowed to be impinged by fire.

Technician Level

Organizations operating at the technician level must have met all requirements for the awareness and operational levels. These organizations must be capable of operating at concrete and steel constructed structures.

The National Incident Management System (NIMS) uses a typing system for all resources that may respond to an incident. There are currently four types of teams eligible to respond to a structural collapse incident. These are:

- TYPE I
- TYPE II
- TYPE III
- TYPE IV

TYPE I teams can conduct safe and effective search and rescue operations at incidents involving collapse or failure of heavy floor, precast concrete, and steel frame construction. Personnel assigned to TYPE I teams must hold the following credentials:

- Trained to the HazMat Technician Level (NFPA 472)
- Comply with NFPA 1006 Technician Level for their organization or area of specialization
- Operations Level for support personnel as outlined in NFPA 1670

TYPE I teams must complete the following training regimens:

- Heavy floor construction
- Pre-cast concrete construction
- Steel frame construction
- High-angle rope rescue
- Confined space rescue (permit required)
- Mass transportation rescue

TYPE II teams can conduct safe and effective search and rescue operations at structural incidents involving the collapse or failure of heavy wall construction. Personnel assigned to TYPE II teams must hold the following credentials:

- Trained to the HazMat First Responder Operational Level (NFPA 472)
- Comply with AHJ requirements
- Operations Level for support personnel as outlined in NFPA 1670

TYPE II teams must complete the following training regimens:

- Heavy wall construction
- High-angle rope rescue (not including high line systems)
- Confined space (no permit required)
- Trench and excavation rescue

TYPE III teams can conduct safe and effective search and rescue operations at structure collapse incidents involving the collapse or failure of light frame construction. Personnel assigned to TYPE III teams must have the following credentials:

- Trained to the HazMat First Responder Operations Level (NFPA 472)
- Comply with the AHJ requirements
- Operations Level for support personnel as outlined in NFPA 1670

TYPE III teams must complete the following training regimen:

- Trained for light frame construction
- Trained for low-angle rope rescue

TYPE IV teams can conduct safe and effective search and rescue operations at incidents involving nonstructural entrapments with minimal removal of debris and building contents. Personnel assigned to TYPE IV teams must have the following credentials:

- Trained to HazMat First Responder Awareness Level (NFPA 472)
- Comply with AHJ requirements
- Awareness Level for support personnel as outlined in NFPA 1670

TYPE IV teams must complete the following training regimen:

- Trained for surface rescue and nonstructural entrapment in noncollapsed structures

Fire departments who wish to have a special operations team or incident management team (IMT) will need to consider the cost and time it will take for personnel to complete the training necessary to attain the credentials. A department must be staffed at least four deep in every position and category to account for 24/7 coverage. The better and more complete solution might be to amalgamate teams among several departments.

Tiered Response

Municipalities must examine their resource capabilities and determine at what level they can support local potential incidents. For many departments it is not practicable to train and certify their members at the technician level. Therefore many departments operate under a tiered response system. With this system, certain members of the department are certified at technician level and are supported by other members of the department certified at the awareness and operations levels. For example, engine companies may be certified at awareness level, truck companies at operations level, and rescue companies at technician level.

A regional response involving multiple departments is another method for ensuring that adequate resources of personnel are available for responses involving collapsed structures.

Special Concerns

Building collapses can be complex events. The original structural instability is even more complicated if the incident becomes a mass casualty event. If terrorism is involved it is a crime scene as well. Hazardous material issues are always a concern. Responses to such events depend more heavily on planning and logistics than operations. Logistics has the critical mission of ordering a large amount of resources unusual for fire departments. Some of these include:

- Cranes and other heavy equipment
- Cutting torches and gases
- Communications resources
- Command post facilities large enough to house all the command functions
- Fencing
- Fleet of trucks for hauling away material

The person in the planning section chief position must be experienced, competent, and have a thorough knowledge of building processes and materials. The planning chief must meet with technical specialists and be able to speak their language. S/he must be able to support operational plans for extrication attempts. Although the incident commander has ultimate go–no-go responsibility, most will rely heavily on the opinion of the planning chief.

US&R TEAMS

A US&R team contains sixty-two trained and certified members. Each member is equipped for large and complex search and rescue operations. The areas of responsibility of the US&R team include command, medical, structural specialists (including engineers), rigging and heavy equipment, rescue, hazardous materials specialists, technical information specialists, safety officers, and others. It will take some time, however, before the US&R teams reach the scene. It is possible to request advance teams, usually a smaller command cell consisting of 4-6 personnel of the main team, and also crews, 11-22 personnel under the direction of a crew leader with a specific mission to deploy. It is certain that early interaction between the US&R staff and the local command cell is important to ensure that cooperation is carried out effectively.

SCHOOL OF HARD KNOCKS

I have responded to many structural collapses over the course of my career. One in particular involved an early twentieth-century steel and masonry building that was being demolished. The crews were working during a weekend, and when the workers removed a primary support column a major collapse occurred, trapping one worker under a steel beam and masonry. The units stabilized the area around the extrication site and the extrication went smoothly. Coordination among all the rescue disciplines went well, allowing paramedics to check the victim periodically.

The operation was assumed to be a success—until the units lifted the steel beam and the victim died instantly, bleeding out internally from crush syndrome. The lesson? Understand crush injuries and have technical rescue certified paramedics involved in the rescue process from the beginning. Then a victim won't be lost after eight hours of effort and the money and resources involved will have been well spent.

FIRST RESPONSE

Special Operations Events

Special operations means just what the term implies. These types of responses are more unique, more technically challenging, and more complex than everyday responses. It takes many years of study, training, and maturity to be an effective member of a special operations unit. Special operations involves many different facets: Units respond to HazMat weapons of mass destruction (WMD) incidents, trench and excavation failures, water borne emergencies, structural collapses, high-angle rescues and mass casualty events.

The members of special operations units must possess many skills beyond competency with tools and book knowledge gained in credentialed training. They must also be carpenters, welders, plumbers, and well versed in rigging. They should all be paramedics. Special operations is to firefighting what the Green Berets are to the military: A team that can not only be self-sufficient but can be split into very effective subcomponents when augmented by skilled and experienced personnel from other units.

The officers and incident commanders for these units and incidents must possess the credentials, skills, and experience to lead such unique and talented teams of responders.

Some departments strategize for the long term when they create a special operations division. The department funds the training, tools, and apparatus for the immediate future, but invests in the long term as well. They will support not only the immediate needs for response to complex these incidents, but they ensure a steady supply of personnel to move into special operations positions through an apprenticeship regimen. Two departments with which I am familiar do this. The cities of Philadelphia and New York have heavy rescue companies that perform special operations responses. These companies are augmented by squad companies staffed with equally trained and credentialed members. The squad companies are usually engine companies that respond to the usual auto accidents, fires, and medical emergencies, but they also have additional equipment on board to augment or support heavy rescue companies when a special operations call comes in. The selection process involved for these departments is unique, involving:

- Interviews
- Testing
- Probationary periods during which potential members are evaluated by the officers
- Constant monitoring after assignment to ensure that members are completing the training necessary for credentialing

This type of system ensures the success of the succession plan. The low probability, but high consequence, of all special operations responses demands that the units be ready to respond and have the personnel and equipment necessary. The succession plan, a process for identifying and developing internal people with the potential to fill key leadership positions in the unit, also involves the officer cadre. Special operations is one of the few units and divisions that cannot be staffed by just anyone nor can it be commanded by just anyone. The personnel in command positions must be knowledgeable, motivated, experienced, and mature enough to command these units in battle.

The responses to the Murrah Building, various weather events, and the World Trade Center are testimony that the health and future of the special operations responders in the United States is second to none.

ON SCENE

You are the responding incident commander at a reported building collapse of a five-story building. While you are en route, the first arriving units report what appears to be a major collapse in a facility that houses elderly patients undergoing rehabilitation. The time is 2130 hours. The temperature is 42°F. The units are reporting that some people can be seen at the top of the pile of debris and that they appear disoriented. The units are reporting that they are going to attempt to mount a rescue attempt by climbing on the pile and are requesting additional companies for support.

1. What orders must you give immediately to protect your people?
2. What information do you need before you can develop strategies and tactics?
3. What additional resources must you request?

Summary

Structural collapse events severely tax or overwhelm most departments. Many times the event is complex, complicated by the additional concerns of mass casualty, HazMat, and secondary collapses. The first arriving units must immediately perform reconnaissance and structural triage, remove apparatus, and begin to request resources. They should not enter the damaged structures or begin work in the debris pile; if they are not certified then safety will certainly be compromised. This is not a seat-of-your-pants, common sense-only type of response. Collapse scenes are extremely dangerous and making accurate on-scene evaluations and assumptions require technical expertise.

Many factors affect the safety and success of operations, one important factor being whether or not there is enough command assistance. Competent personnel must fill the general staff and command staff positions. The best individual for the job might not be from your own department. When communicating, use only terms consistent with the NIMS protocols. Do not use local or colloquial terms because non-local responders might not be familiar with those terms. The correct staffing of the positions of planning and logistics is extremely important to the outcome of the event.

Review Questions

Discuss the responsibilities of the first arriving units.

1. Why is structural triage so important?
2. Why are planning and logistics more important than operations?
3. List some of the technical specialists that may be needed.
4. How long does it take for US&R to arrive at your scene?
5. How will PIAs and lessons-learned help the fire service to improve?
6. Discuss structural collapses to which you have responded.

Suggested Reading

City of Oklahoma City, Oklahoma. April 16, 1996. *Alfred P. Murrah Federal Building Bombing, April 19, 1995, Final Report.*
Introduction to NIMS for the Fire Service, 2002. Emmitsburg, MD: National Fire Academy.

NFPA 1670. 2004. *Standards on Operations and Training for Technical Rescue Incidents.* Quincy, MA: National Fire Protection Association.
NFPA 1700. 2006. *Fire Safety and Emergency Symbols.* Quincy, MA: National Fire Protection Association.

INDEX